Forest Water Ecosystems

Developments in Hydrobiology 13

Series editor
H. J. Dumont

DR W. JUNK PUBLISHERS THE HAGUE-BOSTON-LONDON 1983

Forest Water Ecosystems

Nordic symposium on forest water ecosystems held at Färna, Central Sweden, September 28–October 2, 1981

Edited by
Curt Forsberg and Jan-Åke Johansson

Reprinted from *Hydrobiologia,* vol. 101, no. 1/2 (1983)

DR W. JUNK PUBLISHERS THE HAGUE-BOSTON-LONDON 1983

Distributors:

for the United States and Canada

Kluwer Boston, Inc.
190 Old Derby Street
Hingham, MA 02043
U.S.A.

for all other countries

Kluwer Academic Publishers Group
Distribution Center
P.O. Box 322
3300 AH Dordrecht
The Netherlands

Library of Congress Cataloging in Publication Data

Nordic Sympsium on Forest Water Ecosystems (1981 :
Färna, Sweden)
Forest water ecosystems.

(Developments in hydrobiology ; v. 13)
"Reprinted from Hydrobiologia, vol. 101: 1-2."
1. Lake ecology--Congresses. I. Forsberg, Curt,
1931- . II. Johansson, Jan-Åke. III. Title.
IV. Series.
QH541.5.L3N67 1981 574.5'2632 83-8418

ISBN-13: 978-94-009-7286-5 e-ISBN-13: 978-94-009-7284-1
DOI: 10.1007/978-94-009-7284-1

Cover design: Max Velthuijs

Preface

'A lake is the landscape's most beautiful and expressive feature. It is earth's eye; looking into which the beholder measures the depth of his own nature. The fluviatile trees next the shore are the slender eyelashes which fringe it, and the wooded hills and cliffs around are its overhanging brows.'

'A field of water betrays the spirit that is in the air. It is continually receiving new life and motion from above.'

'Nothing so fair, so pure, and at the same time so large, as a lake, perchance, lies on the surface of the earth. Sky water. It needs no fence. Nations come and go without defiling it. It is a mirror which no stone can crack, whose quicksilver will never wear off, whose gilding Nature continually repairs; no storms, no dust, can dim its surface ever fresh; – a mirror in which all impurity presented to its sinks, swept and dusted by the sun's hazy brush, – this the light dust-cloth, – which retains no breath that is breathed on it, but sends its own to float as clouds high above its surface, and be reflected in its bosom still.'

(Henry David Thoreau: 'Walden; or Life in the Woods'. Fourth printing, April 1953, Rinehart & Company, New York).

As is well known the conditions for life in forest lakes have changed during the short time of man's activities. Drainage, fertilization, large clear-cutting and acidification are examples of influences which were unfamiliar to Thoreau when he expressed his feeling for the forest lake about 125 years ago.

Environmental stress changes the structure and function of the ecosystems of these lakes. In order to understand the further development of these systems we need more basic knowledge. A Nordic symposium held at Färna, central Sweden in 1981 aimed to contribute to an increased understanding of the function of ecosystems of forest waters. A selected number of papers from this symposium are presented below.

Curt Forsberg
Jan-Åke Johansson

Contents

Input–output budgets at Langtjern, a small acidified lake in southern Norway

Richard F. Wright
Norwegian Institute for Water Research, P.O. Box 333, Blindern, Oslo 3, Norway

Keywords: forest water ecosystems, acid precipitation, Norway, chemical budgets, sulfate

Abstract

Precipitation and streamwater volume and chemical composition have been measured since 1974 at Langtjern, a small, acid (pH 4.6–4.8) lake on granitic-gneissic bedrock in coniferous forest located ca. 100 km north of Oslo, Norway. The area receives acid precipitation (weighted average pH 4.28).

The 7-year input–output budgets for major ions at two terrestrial subcatchments indicate that for Na, K, SO_4 and Cl outputs approximately equal inputs, for H^+, NH_4 and NO_3 outputs are much less than inputs, and for Ca, Mg and Al outputs greatly exceed inputs. The sulfate budgets (which include estimated dry deposit) indicate that the terrestrial catchment retains about 20% of the incoming sulfate, perhaps due to absorption in the soil, plant uptake, reduction and storage in peaty areas or reduction and release of H_2S to the atmosphere.

The budgets for Langtjern lake itself indicate that for most components output equals inputs to within 10%, i.e. these compounds simply pass through the lake. For H^+, and possibly NH_4 and NO_3, inputs exceed outputs. Because gaseous phases are not measured the N budgets are uncertain. A mechanism that leads to 'retention' of both H^+ and SO_4 is sulfate reduction and incorporation of sulfides in the lake sediments. Such has been documented in the experimentally-acidified Lake 223, Experimental Lakes Area, Ontario, Canada. Although there is no evidence suggesting the development of anoxic bottom waters at Langtjern, such reduction might occur at the water–sediment interface and in the sediments.

The budgets for the pollutant components H^+ and SO_4 at Langtjern differ substantially from those at the relatively unaffected Lake 239, in the Experimental Lakes Area.

Introduction

Input–output budgets offer a useful tool for investigating the structure and function of catchment and lake ecosystems (Bormann & Likens 1967). Changes both external and internal to the ecosystem are often reflected in short- or long-term alterations in input–output relationships.

Such budgets have been used to study the effects of acid precipitation on terrestrial and lake ecosystems in Norway (Wright & Johannessen 1980).

At Langtjern a 7-year record of inputs and outputs to a lake and its terrestrial catchment is available.

Langtjern is a 0.23 km² lake located about 100 km north of Oslo. Its 4.59 km² catchment is comprised of sparse coniferous forest on podzolic soils and granitic-gneissic bedrock (Figs. 1 and 2). The area receives acid precipitation with weighted average pH of 4.28. Langtjern is acidic (pH 4.6–4.8) and lost its natural population of brown trout in the 1960's probably due to recruitment failure as a result of the acidification of the spawning streams.

Since late 1973 Langtjern has been the site of studies of precipitation, streamwater and lakewater chemistry. From 1973–1978 these studies formed part of the Norwegian SNSF-project (Acid precipitation – effects on forest and fish), in 1979 they were

Hydrobiologia 101, 1–12 (1983).

Fig. 1. Photo of Langtjern looking north.

continued by the Norwegian Institute for Water Research, and in 1980 Langtjern became one of 5 field research areas as part of the Norwegian national environmental monitoring program.

A detailed description of Langtjern and the analytical data are given by Henriksen (1976), Henriksen & Wright (1977), and Wright & Henriksen (1980). Physical, hydrological and chemical characteristics of Langtjern are given in Tables 1 and 2.

Methods

At Langtjern the outlet is equipped with a weir and water-level recorder (Fig. 2) installed and operated by the Norwegian Electricity and Water Resources Board (NVE). Daily samples of precipitation are collected at Gulsvik, 8 km west of Langtjern, and chemical analyses are conducted at the Norwegian Institute for Air Research (NILU). Samples of inflowing and outflowing streams are collected with at least weekly frequency and analyzed at the Norwegian Institute for Water Research (NIVA). Together these data provide the basis for construction of hydrologic and chemical budgets for the terrestrial catchment and lake itself.

Analytical methods

Analytical methods for precipitation and streamwater samples at NILU and NIVA, respectively, are similar. pH is measured potentiometrically, Na, K, Ca, Mg by atomic absorption spectrophotometry, and Al, NH_4, Cl, SO_4 and NO_3 colorimetrically. The water samples are not filtered.

Hydrologic budgets

For the terrestrial catchment the hydrologic budget is comprised of precipitation inputs and outputs via runoff (both streamwater and diffuse sources) and evapotranspiration. Ideally all three components are measured directly. In practice it is difficult

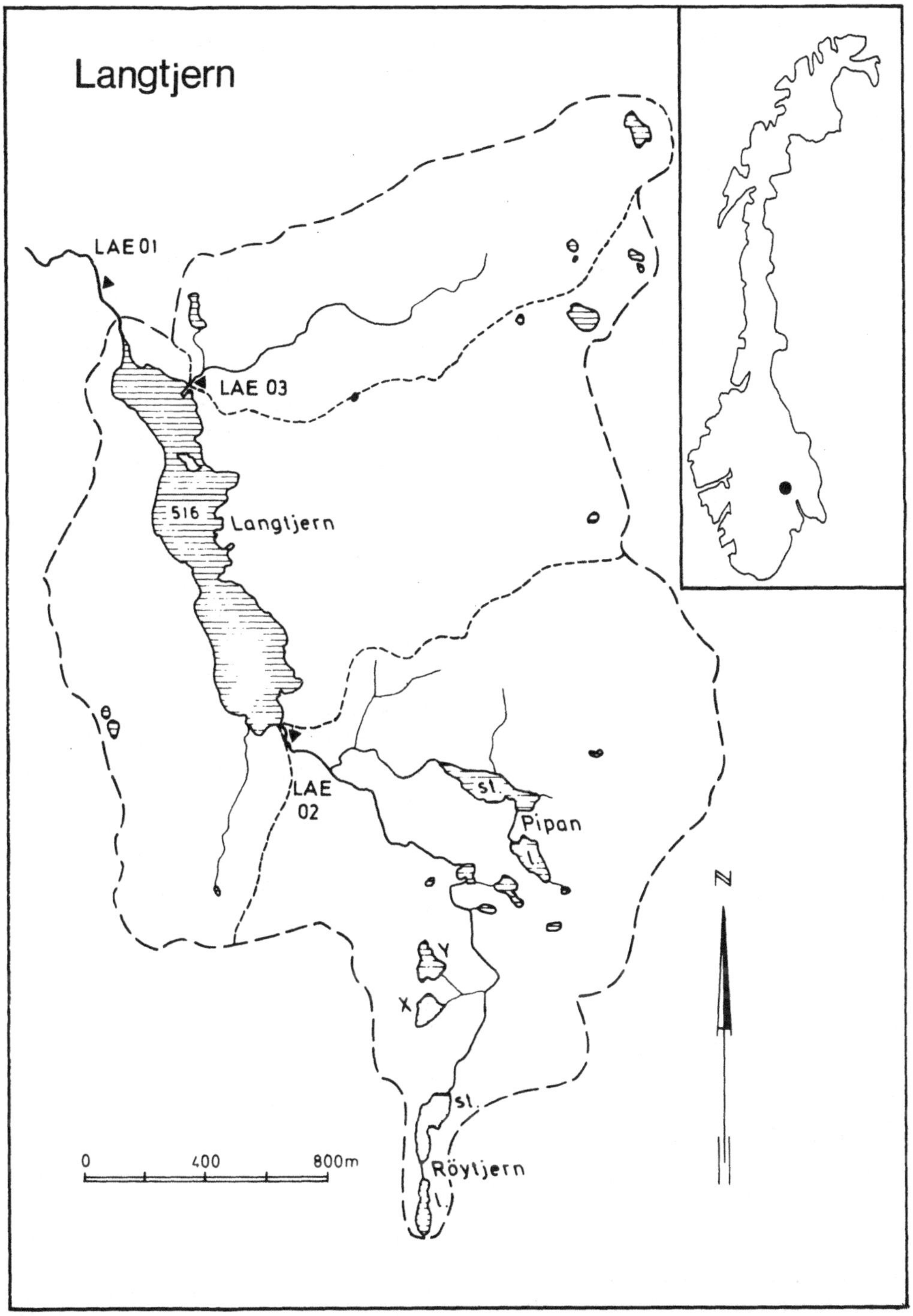

Fig. 2. Map of Langtjern and its catchment showing the outlet stream (LAE01), two inflowing streams and their subcatchments (LAE02, LAE03) and the sampling site in the north basin of the lake.

Table 1. Physical and hydrological data for Langtjern and its catchment.

Lake area		0.23 km²
Terrestrial catchment		
subcatchment, stream LAE02	1.81 km²	
subcatchment, stream LAE03	0.75 km²	
remaining areas, diffuse	2.00 km²	
Total		4.56 km²
Lake and catchment, outlet LAE01		4.79 km²
Lake elevation above sealevel		516 m
Maximum elevation in catchment		720 m
Lake		
maximum depth		12 m
average depth		2.0 m
volume		0.56×10^6 m³
Theoretical water retention time (lake volume/mean outlet discharge)		74 days

Table 2. Volume-weighted average concentrations (1974–80) of major ions in precipitation and streamwater at Langtjern. Units: $\mu eq \cdot l^{-1}$.

	Precipitation (total)	Inflowing streams LAE02	Inflowing streams LAE03	Outlet LAE01
H_2O (mm)	876	576	576	576
H^+	53	25	30	20
(pH)	(4.28)	(4.60)	(4.52)	(4.69)
Na	8^a	21^c	22^c	26^c
K	3^b	3	2	5
Ca	8	61	48	61
Mg	3	16	17	19
Al^d	–	25	27	26
NH_4	27	–	–	–
SO_4	57	85	82	82
Cl	9	21	21	18
NO_3	29	2	1	2
Σ^+	102	126 + Al	119 + Al	131 + Al
Σ^-	95	108	104	102

[a] 1974–76 estimated from Cl and Na/Cl in seawater.
[b] For years 1974, 1975, 1976 and 1980 only.
[c] For years 1975–80.
[d] Measured as total dissolved Al.

if not impossible to directly measure evapotranspiration, and accurate measurement of precipitation inputs to an entire catchment is also problematic. Streamwater runoff, on the other hand, can be measured to within a few percent with a weir and water-level recorder.

Evapotranspiration is estimated at 300 mm · year^{-1} on the basis of maps of potential evapotranspiration given by Forsman (1976). Actual evapotranspiration varies from year-to-year, and the estimate is probably no better than ±50 mm · year^{-1}.

Inflow stream discharge has been gauged directly using temporary weirs for a portion of the 7-year period 1974–1980, but the data are incomplete. A satisfactory estimate can also be obtained using the lake outlet discharge (which has been measured continuously since late 1973) and a simple model for the lake hydrology (termed the BV3 model, Wright & Henriksen 1980). Runoff from areas not drained by one of the two streams is assumed proportional to that of stream 03. Such diffuse sources (such as empheral streams and runoff) are difficult to measure directly. Precipitation volume is measured at Gulsvik, but the reported volume is clearly less than actual precipitation volume at Langtjern. For some years measured runoff at the lake outlet exceeded measured precipitation, a hydrologic paradox.

Undermeasurement of precipitation volume can occur because the gauge placement is not representative for the catchment as a whole and because of less-than-100% collection efficiency of the gauge. At Langtjern both factors probably are important. The precipitation station at Gulsvik is not only 8 km distant from the catchment, but it is located in a valley at only ca. 200 m above sea level whereas Langtjern is 516 m, and the top of the catchment 720 m. Because of orographic effects precipitation

Table 3. Calculation of gauge factor for precipitation for the years 1974–80 using discharge data for the outflow stream (LAE01), estimated evapotranspiration and measured precipitation volume. Units: mm · y^{-1}.

	Precipitation measured	Precipitation total	Precipitation gauge factor	Evapotranspiration	Runoff
1974	783	936	1.20	300	636
1975	560	819	1.46	300	519
1976	641	639	1.00	300	339
1977	683	1 047	1.53	300	747
1978	608	929	1.53	300	629
1979	790	902	1.14	300	602
1980	667	860	1.29	300	560
mean 1974–80	676	876	1.30	300	576

up at Langtjern is likely to be greater than at Gulsvik.

Precipitation volume is thus obtained by difference from the hydrologic budget calculated for each year.

$$P = R + ET$$

where P is precipitation, R runoff and ET evapotranspiration. The results for each of the 7 years for the area drained by the outlet stream LAE01 (lake plus catchment, 4.79 km^2) are given in Table 3. The ratio between total estimated and measured precipitation is termed 'gauge factor' and is the factor by which the measured precipitation inputs of water and chemical species are multiplied to obtain total wet inputs.

Chemical budgets

Atmospheric inputs

Inputs of chemical components include inputs in wet precipitation and dry deposition. The latter comprises gravitational settling of particles and gases and impaction on vegetational and other surfaces.

Dry deposition inputs to an entire catchment are difficult if not impossible to measure (Wright & Johannessen 1980). For sulfur an estimate can be obtained from SO_2 gas and SO_4 particle concentrations in air and representative deposition velocities. At the precipitation station at Gulsvik average daily SO_2 and SO_4 levels for the 14-month period were 1.2 $\mu g\ SO_2 \cdot m^{-3}$ and 2.5 $\mu g\ SO_4 \cdot m^{-3}$, respectively. The deposition velocity for SO_2 probably lies in the range 0.5–1.0 $cm \cdot s^{-1}$ and for SO_4 0.1–0.7 $cm \cdot s^{-1}$ (Joranger *et al.* 1980). Using 0.8 and 0.2, respectively, the dry deposition amounts to 13 $keq\ SO_4 \cdot km^{-2} \cdot year^{-1}$. This amount is assumed to apply for the entire 7-year period at Langtjern, and the uncertainties in the estimate are at least 50%. Dry deposition of H^+ is assumed to be an equivalent amount as for SO_4, based on the rationale that all sulfur entering as dry deposit is oxidized and hydrolyzed to SO_4 producing an equivalent amount of H^+ in the process.

For components whose major source is seawater spray, dry deposition can be estimated from the chloride budget, given the assumption that total inputs of chloride equal output in runoff and that the catchment is neither a net source nor sink for chloride. Dry deposition of other marine salts is then assumed proportional to the ionic ratio to chloride in seawater (Wright & Johannessen 1980).

At present there are no data available for estimating dry deposition of nitrogen compounds and dry deposition of terrestrially-derived dusts.

For wet deposition the measured inputs are scaled by the gauge factor obtained from the hydrologic budgets. This assumes that uncollected precipitation has the same chemical composition as weighted average annual wet precipitation.

Runoff outputs

All outputs are assumed to occur as dissolved in streamwater. Particulates not included in the chemical analyses (samples are not filtered) and gaseous exports are assumed to be negligible.

Mass export from the 2 streams and outlet are calculated using the daily discharge values (from the BV3 model for the 2 inflowing streams, measured directly for the outlet) and the weekly concentration values. Since discharge is measured more frequently than concentration, export is calculated for those days for which chemical data exist and then extrapolated to other days using a discharge weighting procedure.

In effect this procedure calculates the discharge-weighted concentration for days on which chemical measurements are available, then multiplies this value by the total discharge over the entire period to obtain export (see Wright & Henriksen 1980 for details).

Influx to the lake from diffuse sources is assumed to be proportional to that from stream 03.

Results

Hydrologic budgets

The Langtjern data permit the construction of hydrologic budgets for the terrestrial subcatchments drained by each of the 2 major inflowing streams, for the lake itself, and for the lake and its entire catchment.

The hydrologic budgets for the terrestrial catchment and for Langtjern itself for each of the 7 years

Table 4. Hydrologic budgets for the terrestrial catchment (4.56 km²) and for Langtjern lake itself (0.23 km²) for the period 1974–80.

Year	Terrestrial catchment			Lake				Water retention time (years)
	in	out		in		out		
	Precipitation 10^6 m³	Runoff 10^6 m³	Evapotranspiration 10^6 m³	Precipitation 10^6 m³	Inflow 10^6 m³	Evaporation 10^6 m³	Outflow 10^6 m³	
1974	4.27	2.77	1.37	0.18	2.77	0.06	3.05	0.18
1975	3.73	2.43	1.37	0.13	2.43	0.06	2.49	0.22
1976	2.91	1.48	1.37	0.15	1.48	0.06	1.62	0.35
1977	4.77	3.36	1.37	0.16	3.36	0.06	3.58	0.16
1978	4.24	2.89	1.37	0.14	2.89	0.06	3.01	0.19
1979	4.11	2.83	1.37	0.18	2.83	0.06	2.88	0.19
1980	3.92	2.62	1.37	0.15	2.62	0.06	2.68	0.21
mean 1974–80	3.99	2.63	1.37	0.16	2.63	0.06	2.75	0.20

Table 5. Input–output budgets for the subcatchments drained by inflowing streams 02 and 03 and for the outflowing stream 01 at Langtjern. Values are averages for the 7-year period 1974–80. Units: Keq · km^{-2} · y^{-1}. Areas are 01 4.79 km², 02 1.81 km², and 03 0.75 km².

	In				02		03		01	
	meas. ppt	total[a]	dry deposition[b]	Σ in	out	out/in	out	out/in	out	out/in
H_2O (mm)	676	876			576	–	576		576	
H^+	35.7	46.3	13	59	13.9	0.24	17.1	0.29	11.7	0.20
Na	5.2	6.7	3.4	10.1	14.9	1.48	16.6	1.64	15	1.49
K	2.7	3.5	0	3.5	2.7	0.77	2.4	0.69	2.7	0.77
Ca	5.3	6.9	0.1	7.0	38	5.4	32	4.6	35	5.0
Mg	1.9	2.5	0.8	3.3	10.7	3.2	12.4	3.8	11	3.3
Al	–	–	–	–	15[c]	–	16[c]	–	15[c]	–
SO_4	38.4	49.8	13	63	49	0.78	50	0.79	47	0.75
Cl	5.9	7.6	4.0	10.6	10.6	1.00	11.1	1.05	10.6	1.00
NO_3	19.5	25.3			1.1		0.9		1.4	

[a] Total = measured × gauge factor of 1.30.
[b] Dry deposition estimated as follows:
SO_4 and H^+: from SO_2 gas, 0.6 μg SO_2-S · m^{-3} and deposition velocity 0.8 cm s^{-1} gives 9.5 keq · km^{-2} · y^{-1}.
from SO_4 particles, 0.8 μg SO_4-S · m^{-3} and deposition velocity 0.2 cm s^{-1} gives 3.2 keq · km^{-2} · y^{-1}.
Cl: dry deposition = Cl output – Cl wet inputs at LAE01 = 4.0 keq · km^{-2} · y^{-1}.
Na: dry deposition = 0.86 Cl dry deposition = 3.4 keq · km^{-2} · y^{-1}.
Mg: dry deposition = 0.20 Cl dry deposition = 0.8 keq · km^{-2} · y^{-1}.
Ca: dry deposition = 0.04 Cl dry deposition = 0.1 keq · km^{-2} · y^{-1}.
[c] Measured as total dissolved Al.

reveal that annual water flux through the catchment and lake can vary by about a factor of 2 from year-to-year (Table 4). Because the runoff from the 2 streams LAE02 and LAEO03 is estimated, the calculated budgets for each will be nearly identical to those given in Table 4 in the units mm · y^{-1}. The calculated average annual flux of water at Langtjern indicates that about 1/3 of incoming precipitation returns to the atmosphere as evapotranspiration and 2/3 leaves as runoff. Because the lake occupies only 5% of the entire catchment, precipitation directly on the lake and evaporation from the lake surface represent only a minor fraction of the hydrologic budget for the lake. For the same reason

Table 6. Input–output budgets (1974–80) for Langtjern lake itself (0.23 km²). Units: Keq · y^{-1}.

	Inputs			Outputs	Input–output		Mass in lake water
	precipitation	inflow	Σ	outflow	gain (+) loss (–)	% of input	
H_2O (10^6 m^3)	0.16	2.62	2.78	2.75			0.56
H^+	11	72	83	56	+27	+33	11
Na	2	72	74	72	+ 2	+ 3	15
K	0	12	12	13	– 1	– 8	3
Ca	2	157	159	169	– 10	– 6	34
Mg	1	52	53	53	0	0	11
Al	0	71	71	73	– 2	– 3	15
SO_4	11	227	238	225	+13	+ 5	46
Cl	2	49	51	51	0	0	10
NO_3	6	5	11	7	+ 4	+36	1

average water retention time at Langtjern is only 2–4 months (mean 74 days for 1974–80).

Chemical budgets

The chemical budgets for the terrestrial subcatchments drained by streams 02 and 03 as well as the budgets for the entire catchment and lake show that for Na, K, SO_4 and Cl outputs approximately equal inputs, for H^+, NH_4 and NO_3 outputs are much less than inputs and for Ca, Mg and Al outputs greatly exceed inputs (Table 5). This general pattern is exhibited both by the yearly budgets as well as the 7-year mean. The long-term data from Langtjern thus confirm the conclusions drawn from previous budget calculations for Langtjern presented by Gjessing *et al.* (1976), Henriksen (1976), Henriksen & Wright (1977) and Wright & Johannessen (1980).

These budget trends are relatively insensitive to the method of estimating dry deposition and to the estimate of precipitation gauge factor.

The budgets for the two subcatchments differ slightly due to minor differences in water chemistry. Stream 02 (the larger) has on the average higher Ca and lower H^+ concentrations than stream 03 (Table 2). The hydrologic budgets are similar, thus these differences in concentrations are reflected in commensurate differences in the Ca and H^+ budgets. The slightly different water chemistry is probably due to the presence of somewhat more-easily weathered ground moraine and bedrock in the subcatchment drained by stream 02.

For Langtjern lake itself outputs equal inputs to within 10% except for H^+ and NO_3 (Table 6). For these ions outputs are less than inputs; the lake is a net sink. The lake budgets are essentially independent of the estimates used for dry deposit and gauge factor, because the lake surface is small relative to catchment area (ca. 1:20) and thus most of the inputs come in runoff. The greatest uncertainty lies in the estimate for the chemical composition of runoff from diffuse sources, the parts of the catchment not drained by either of the two major inflowing streams. This diffuse runoff was assumed to have the same chemical composition as stream 03.

Discussion

Precipitation, streamwater and lakewater chemistry at Langtjern is dominated by strong acids, particularly sulfate, the major anion. Langtjern is thus highly affected by chemical components such as H^+ and SO_4. It is illustrative to compare this acidified system with a nonacidified but otherwise similar lake and catchment system. Lake 239 (Rawson Lake) in the Experimental Lakes Area, northwestern Ontario, provides such a reference. Schindler *et al.* (1976) have presented comparable input–output budgets for this lake. Lake 239 is located in mixed coniferous forests on granitic bedrock. The lake is somewhat larger than Langtjern (0.54 km² vs 0.23 km²) and has a somewhat smaller terrestrial catchment (3.42 km² vs 4.56 km²). Both lakes are dimictic and oligotrophic. Langtjern is highly colored and has a very short water retention time as compared with Lake 239 (0.2 y vs 5.3 y).

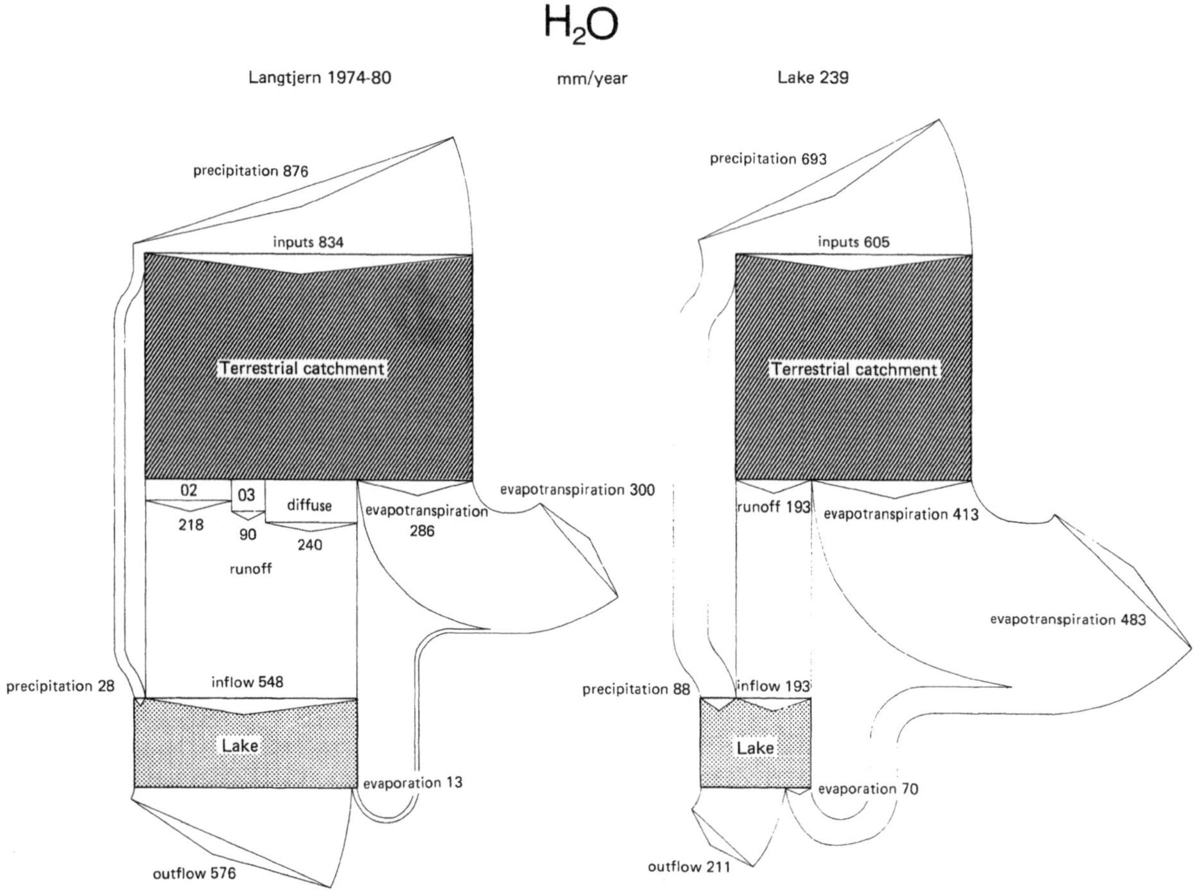

Fig. 3. Mean annual flux of water for the period 1974–80 at Langtjern and its terrestrial catchment (4.79 km²) and for 1972 at Lake 239, ELA (3.96 km²) (Schindler *et al.* 1976). Units: mm · y^{-1}.

The hydrologic budgets for Langtjern and Lake 239 differ in several ways (Fig. 3). At Langtjern evapotranspiration is lower relative to Lake 239 reflecting the cooler summer climate. Because Langtjern occupies only 5% of its basin (Lake 239 14%) the hydrologic budget for the lake is dominated by runoff inflow and outlet outflow. At Lake 239 precipitation and evaporation at the lake surface are relatively more important.

The calcium budgets at both lakes are quite similar, reflecting the facts that the flux of water at both sites is similar and that the rate of release of Ca in chemical weathering is similar (both areas are on granitic bedrock). Atmospheric deposition comprises a significant fraction of the input to the terrestrial catchment, outputs in runoff exceed inputs in atmospheric deposition, and neither lake is a major source or sink for calcium (Fig. 4).

The sulfate budgets, on the other hand, are quite different (Fig. 5). The flux of sulfate at Langtjern is about 3 times larger than that at Lake 239; southern Norway receives acid, sulfate-rich precipitation whereas northwestern Ontario lies at the edge of the areas in North America that receive high loadings of acid from the atmosphere. Sulfate also appears to be relatively conservative in that outputs are nearly equal to inputs in both the terrestrial catchments and lakes.

At Langtjern about 20% of the incoming sulfate appears to be retained in the terrestrial catchment. Although the magnitude of the amount of sulfate retained is uncertain, neither the uncertainties in the estimates of gauge factor or dry deposition are sufficiently large to reject the hypothesis that sulfate is retained. At Lake 239 the data cover only 1 year, and a longer record is necessary before it can

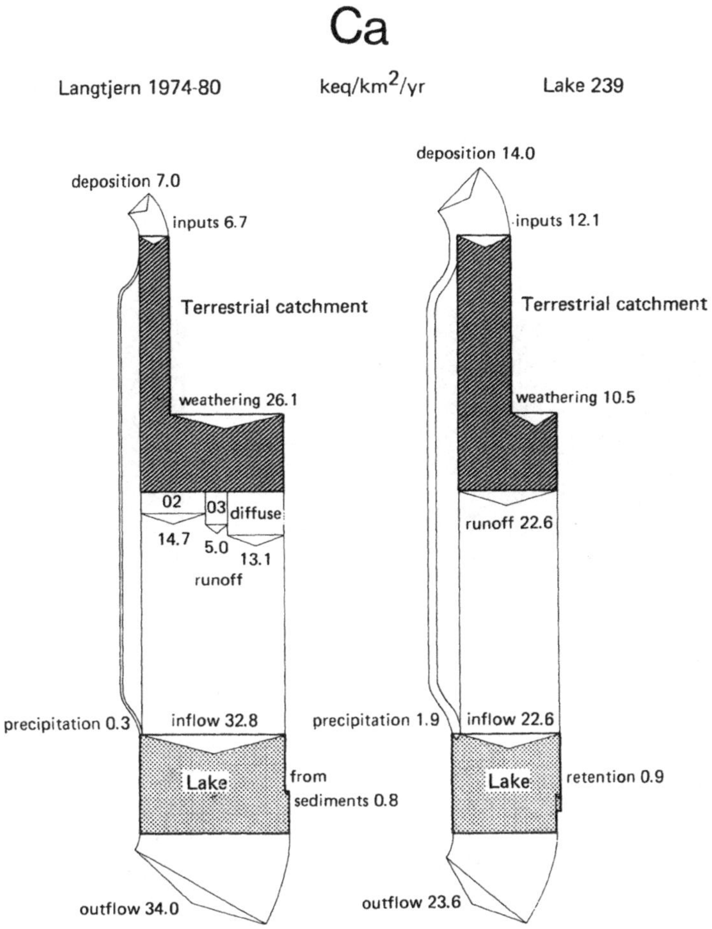

Fig. 4. Mean annual flux of calcium for the period 1974–80 at Langtjern and its terrestrial catchment (4.79 km^2) and for 1972 at Lake 239, ELA (3.96 km^2) (Schindler *et al.* 1976). Units: Keq · km^{-2} · y^{-1}.

be concluded that sulfate is retained or released in the terrestrial catchment. Such data are being collected but have not yet been published.

The sulfate budgets at both Langtjern and Lake 239 suggest that a small fraction is retained in the lakes themselves. This conclusion is uncertain because of the importance of the diffuse runoff in the sulfate budgets.

There are a number of possible processes which could lead to sulfate retention: (1) release of sulfur as H_2S gas to the atmosphere; (2) reduction of sulfate to sulfide and deposition as sulfides in anaerobic soils and lake sediments; (3) uptake and incorporation of sulfur by plants; (4) sulfate absorption in soils and lake sediments.

All of these undoubtedly occur at Langtjern, but the quantitative importance is unknown.

The hydrogen-ion budgets also differ greatly in magnitude at Langtjern relative to Lake 239 (Fig. 6). Lake 239 catchment and lake are bicarbonate systems with pH above 6.0. At Langtjern acidification has proceeded to the stage at which the bicarbonate is replaced by the strong-acid anion sulfate and both stream and lake waters are acidic, pH below 5. The flux of H^+ inputs to the terrestrial catchment at Langtjern is four times that of Lake 239. Although at Langtjern runoff carries only 1/4 the amount of H^+ entering in atmospheric deposition, this amount is sufficient to make the runoff highly acidic (average pH 4.6). The catchment appears to have a limited capacity to neutralize inputs of strong acids. At Lake 239 atmospheric deposition of H^+ is much lower and runoff carries very little H^+; almost all is retained. Runoff has a high

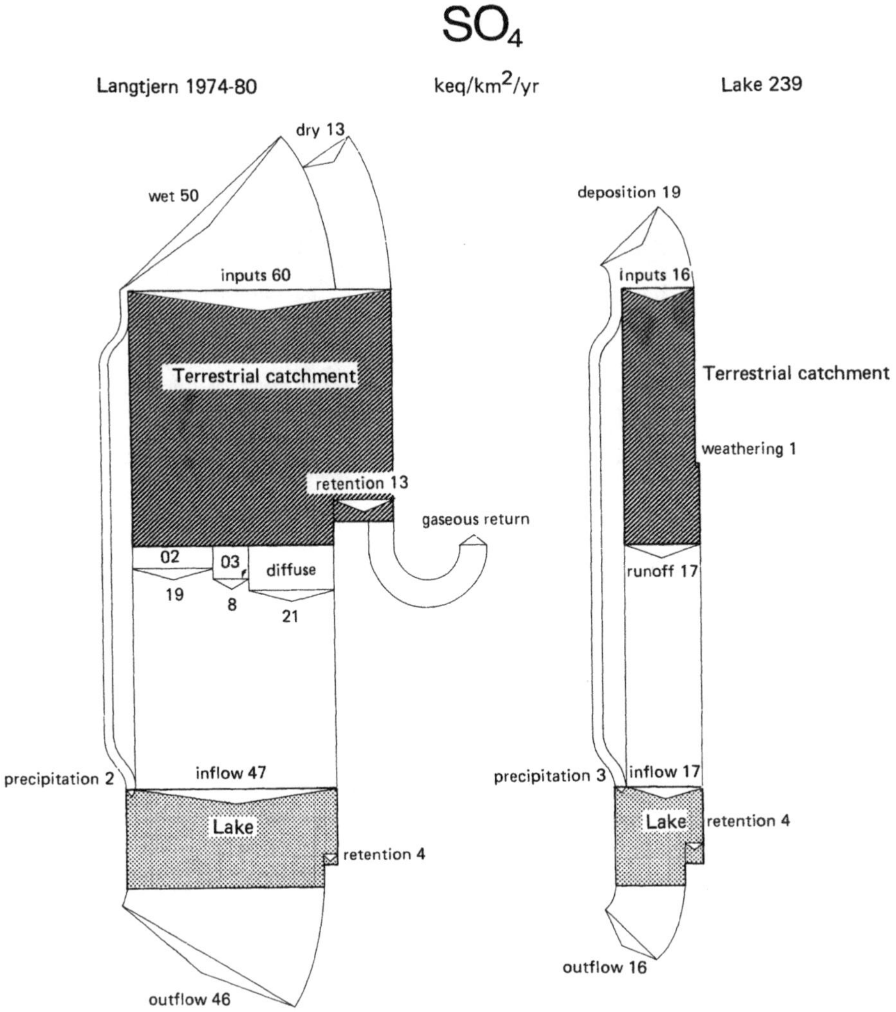

Fig. 5. Mean annual flux of sulfate for the period 1974–80 at Langtjern and its terrestrial catchment (4.79 km²) and for 1972 at Lake 239, ELA (3.96 km²) (Schindler *et al.* 1976). Units: Keq · km^{-2} y^{-1}.

pH, and bicarbonate is present.

An additional fraction of H^+ is apparently retained in the lakes. At Langtjern about 1/3 of the incoming H^+ is retained; at Lake 239 nearly all the H^+ is retained. Again the much larger flux of H^+ at Langtjern means that the lake water and outflow are still low pH, even though the lake retains more H^+ than Lake 239 (5.6 vs 1.7 keq · y^{-1} per km² total catchment and lake area; 117 vs 12 keq · y^{-1} per km² lake surface area).

The budgets for Langtjern lake indicate that within the uncertainties of the measurements outputs equal inputs for all major ions except for H^+, and possibly NH_4 and NO_3. The nitrogen budgets are incomplete because gaseous phases are not included. The sulfate budget indicates a slight retention in the lake, but the uncertainties are such that this may not be significant. In any case there is clearly no retention of sulfate in Langtjern at the scale observed in the artificially-acidified Lake 223 of the ELA (Schindler *et al.* 1981). Here sulfate reduction and incorporation of sulfides in the sediments occurred in the anoxic hypolimnion.

In 1977, the second year of the experiment, approximately 80 keq SO_4 · y^{-1} per km² lake surface were removed from the water and presumably retained in the sediments as iron sulfides. Lake 223, however, is unusual for oligotrophic soft-water

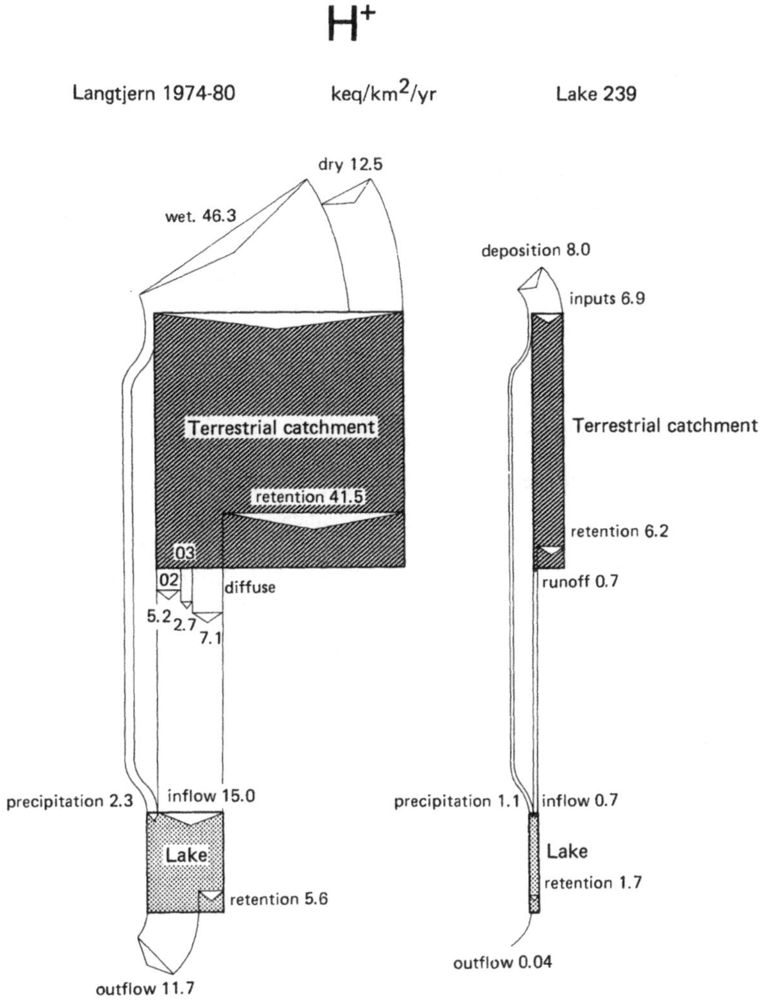

Fig. 6. Mean annual flux of H^+ for the period 1974–80 at Langtjern and its terrestrial catchment (4.79 km²) and for 1972 at Lake 239, ELA (3.96 km²) (Schindler *et al.* 1976). Units: Keq · km^{-2} · y^{-1}.

lakes in that it develops an anoxic hypolimnion in which sulfate reduction is promoted. At Langtjern anoxia does not develop. Sulfate reduction may then occur only in the sediments and rate of reduction governed by diffusion of sulfate into and sulfide out of the pore waters.

The H^+ budgets at both Langtjern and Lake 239 indicate that outflow water is of higher pH than inflowing water due to net retention of H^+ in the lake. In addition to sulfate reduction there are a number of other processes which lead to retention, such as nitrate reduction, primary production, cation exchange and chemical weathering in the sediments. The data are insufficient to estimate the quantitative importance of these processes, although the budgets for the other cations at Langtjern suggest that cation exchange and chemical weathering in the sediments are relatively minor contributors to the H^+ retention.

If the H^+ budget for Langtjern is typical then we have an explanation for the general observation that uppermost, head-water lakes are often lowest in pH and lose their fish populations first (Wright & Johannesen 1980).

Acknowledgements

Hydrochemical research at Langtjern has been financed by the Norwegian SNSF-project (Acid precipitation – effects on forest and fish), the Norwegian Institute for Water Research, and the Norwegian State Pollution Control Authority. The Norwegian Electricity and Water Resources Board supplied discharge data, and the Norwegian Institute for Air Research provided precipitation data. Kolbjørn Sønsteby collected many of the water samples. I thank T. Ahl, R. B. Cook, A. Henriksen and M. Johannessen for helpful discussions.

References

Bormann, F. H. & Likens, G. E., 1967. Nutrient cycling. Science 1955: 424–429.

Forsman, A., 1976. Water balance maps of the Nordic countries. Vannet i Norden 9: 27–41.

Gjessing, E. T., Henriksen, A., Johannessen, M. & Wright, R. F., 1976. Effects of acid precipitation on freshwater chemistry. In: Brække, F. H. (ed.). Impact of Acid Precipitation on Forest and Freshwater Ecosystems in Norway, pp. 64–85. SNSF-project, Ås, Norway.

Henriksen, A., 1976. Chemical investigations of precipitation and river water in the Langtjern basin, southern Norway, 5 May 1973–30 June 1975. Tech. Note 25/76, SNSF-project, 1432 Ås, Norway (in Norwegian).

Henriksen, A. & Wright, R. F., 1977. Effects of acid precipitation on a small acid lake in southern Norway. Nord. Hydrol. 8: 1–10.

Joranger, E., Schaug, J. & Semb, A., 1980. Deposition of air pollutants in Norway. In: Drabløs, D. & Tollan, A. (eds.). Ecological Impact of Acid Precipitation, pp. 120–121. SNSF-project, Ås, Norway.

Schindler, D. W., Newbury, R. W., Beaty, K. G. & Campbell, P., 1976. Natural water and chemical budgets for a small Precambrian lake basin in central Canada. J. Fish. Res. Bd Can. 33: 2526–2543.

Schindler, D. W., Wagemann, R., Cook, R. B., Ruszczynski, T. & Prokopowich, J., 1981. Experimental acidification of Lake 223, Experimental Lakes Area: Background data and the first three years of acidification. Can. J. Fish. aquat. Sci. 37: 342–354.

Wright, R. F. & Henriksen, A., 1980. Hydrologic and chemical studies at Langtjern. Intern. Rept. 71/80, SNSF-project, 1432 Ås, Norway, pp. 108.

Wright, R. F. & Johannessen, M., 1980. Input–output budgets of major ions at gauged catchments in Norway. In: Drabløs, D. & Tollan, A. (eds.). Ecological Impact of Acid Precipitation, pp. 250–251. SNSF-project, Ås, Norway.

Dissolved inorganic substances

A study of mass balance in three small drainage basins

Ulla Maria Calles
Division of Hydrology, University of Uppsala, Västra Ågatan 24, S-752 20 Uppsala, Sweden

Keywords: forest water ecosystems, water chemistry, mass balance, deposition, salts

Abstract

During the International Hydrological Decade (IHD) three drainage basins were selected as being representative for Swedish forest areas. These basins were Velen, Kassjöån and Lappträsket. Although the Representative Basin Project ended in 1975 investigations have been continued in some of the sub-basins. This paper is based on data from three sub-basins within Lappträsket and Kassjöån. Investigations were carried out on chemical composition of runoff and precipitation in order to compute the balance of analysed ions. Annual transports were computed using one analysis per month and monthly mean discharge of water. Since the chloride content of the bedrock is very low all the discharged chloride is regarded as being of atmospheric origin. This assumption has been used for the calculation of balance. Computed balances show a deposition of sulphur that exceeds the amount of sulphur discharged. Cations show a discharge that exceeds deposition in all investigated basins.

Introduction

The present paper is a follow-up of earlier investigations (Andersson & Eriksson 1978, Andersson-Calles & Eriksson 1979). Since these reports were published longer series of data have been obtained which justifies a renewed presentation.

Previous reports were based on data from the International Hydrological Decade's Representative Basin Project. This project ended in 1975 but in some of the sub-basins investigations are still carried out as a part of the FFO-project (Field Research Areas). The responsibility for this project rests with the Swedish Meteorological and Hydrological Institute (SMHI) and the Swedish Geological Survey (SGU). The number of drainage basins referred to in the present paper is smaller than in the previous reports.

The investigated areas

The drainage basins from which data are used are sub-basins within the former representative basins of Lappträsket and Kassjöån (Fig. 1) both of which are regarded as being representative of Swedish forest areas.

Fig. 1. Location of the former representative basins Lappträsket and Kassjöån.

Hydrobiologia 101, 13–18 (1983).

Table 1. Major physiographic data on the investigated drainage basins.

Name	Area (km^2)	Lake area (%)	Bog area (%)
Basin Lappträsket			
Solmyren	27	0.5	33
Vuoddasbäcken	42	0.7	27
Basin Kassjöån			
Lilla Tivsjön	12.8	2.7	8

The bedrock consists mostly of granites and gneisses. Part of the drainage takes place through systems of fractures and fissures that are typical of old hard rock areas. The major part of all the investigated basins are covered with a usually thin deposit of till. Some of the major physiographic data are compiled in Table 1. For a detailed description and characteristics the reader is referred to SMHI publications (Waldenström 1977; Persson 1971).

Methods of sampling and analysis

Precipitation was collected monthly at one station in each of the two major basins and analysed at the Institute of Meteorology, University of Stockholm (MISU). The precipitation chemistry supplied data on the average composition of precipitation, used for estimating the deposition of atmospheric salts. The analysis comprised chloride, sulphate, nitrate, bicarbonate (or strong acid), ammonia, sodium, potassium, calcium, magnesium, pH and electrical conductivity. Water samples were collected on the 15th of every month at all the discharge stations. These samples were also analysed at MISU. Details of the chemical analysis used is found in the Nordic IHD Report No. 3 (Anonymus 1972).

During 1978 no samples were collected at the discharge stations. In 1979 the sampling date was changed to the 1st of each month. It should also be noted that the analyses for 1979 used in the present paper were done at the Royal Institute of Technology (KTH) in Stockholm.

Computation of mass balance

Data on concentration of ions obtained in the analyses were used in combination with values of monthly discharges of water to compute the monthly discharge of dissolved substances. Monthly values were added up to hydrological years using the period from October one year to September next year.

The use of only one sample per month means that a certain bias may occur. This effect will however be small as long as the concentrations are independent of water discharge. The bias in this type of computation has been studied by Eriksson (1974) who compared this method with one using several samples per month. It was found that the difference in total annual amounts amounted to less than 5% and for monthly values was up to 10% with a maximum error occurring during the snow melt period.

The input of dissolved substances from the atmosphere was computed using analyses of monthly samples of precipitation combined with areal means of the amount of precipitation. The contribution from 'dry' deposition is unknown and has been ignored, which means that, e.g., in the case of sulphur the total deposition is likely to be larger particularly since sulphur can also occur in the atmosphere as sulphur dioxide, and this gas can be absorbed directly by vegetation and hence contribute to the total deposition.

The basic idea for calculation of balance of dissolved substances is that the concentration of chloride in igneous rocks is extremely low. The chloride discharged in rivers must therefore come from other sources than weathering of those rocks. Two explanations are possible. Either the chloride consists of airborne sea salts (cf. Eriksson 1952, 1955, 1958, 1959, 1960) or it consists of 'fossil' sea water in the fissures of the bedrock (Andersson-Calles & Eriksson 1979). The latter is only applicable for areas that were covered by the sea at the late stages of the last glaciation. Considering the time passed it is, however, unlikely that it has an effect on discharge of chloride.

The chemical composition of precipitation averaged over the entire period of investigation can be assumed to represent in a relative sense the average total deposition in basins where chloride deposition by precipitation roughly balances discharge. The ratios between the various constituents obtained from long-term averages are thus supposed to apply also for the actual deposition. This assumption is valid for those elements that are present in the atmosphere as airborne particles, but in the case of sulphur the assumption can be misleading since sulphur can also occur in the atmosphere as sulphur dioxide which is a gas that can be absorbed directly by vegetation and hence contribute to the total deposition. Using precipitation chemistry data will most likely underestimate sulphur deposition. The error will be most pronounced in areas that are likely to have high concentrations of sulphur dioxide.

A slight difference between deposition of chloride and discharge of chloride can be observed in Table 2. The reason for this may lie in the way deposition is computed. Concentrations are determined in the monthly samples from chemistry precipitation gauges. The amount of precipitation on the other hand is not determined from these gauges but from precipitation gauges elsewhere in the drainage basin. An areal mean amount is computed from the precipitation gauges and multiplied by the concentration value. Since the design of the two types of gauge differs an error may be introduced. Evaporation is probably somewhat higher in the chemistry precipitation gauges which will give too high concentrations. On the other hand these gauges are less efficient particularly considering snow precipitation. The amounts deposited in these gauges are therefore usually less than the actual deposition.

The small discrepancies between calculated deposition and discharge is probably due to the calculation method and do not contradict the hypothesis that the chloride amounts should be the same. A practical conclusion is that deposition data should be adjusted to such a level that chloride balances at least for elements which occur as particles in the atmosphere. However, such an adjustment does not necessarily correct the total deposition of sulphur to a proper volume since sulphur can be removed from the atmosphere also as a gas.

In general it can be noted that sulphur does not balance; much less is discharged than is deposited. For the different areas the deficit amounts to 25 to 65% of the deposition or 230 to 580 kg · km^{-2} · year^{-1}. Keeping in mind sulphur deposition by direct absorption of sulphur dioxide from the atmosphere the deficit could easily be even larger. Two possible explanations should be considered. One is a steady accumulation of sulphur in soil organic matter. However, such storage changes can not go on for ever and the explanation is thus doubtful. Of

Table 2. Deposition, discharge and balance in the three investigated basins. Unit is kg · km^{-2} · year^{-1}. Mean values for the hydrological years 1969/70–1978/79. Deposition is computed by adjusting deposition by precipitation to such a level that chloride balances.

	S	Cl	NO_3-N	NH_3-N	Na	K	Mg	Ca
Basin Lappträsket								
(Precipitation 621 mm)								
Precipitation	783	253	156	156	283	107	76	384
Solmyren								
(Runoff 348 mm)								
Discharge	684	296	15	12	579	224	413	1 649
Deposition	916	296	183	183	331	125	89	449
Balance	-232	0	-168	-176	+248	+99	+324	+1 200
Vuoddasbäcken								
(Runoff 383 mm)								
Discharge	323	293	21	11	676	159	343	1 095
Deposition	907	293	181	181	328	124	88	445
Balance	-584	0	-160	-170	+348	+35	+255	+650
Basin Kassjöån								
(Precipitation 623 mm)								
Precipitation	774	245	200	206	198	101	66	374
Lilla Tivsjön								
(Runoff 203 mm)								
Discharge	395	218	19	11	334	96	324	1 178
Deposition	669	218	178	183	176	90	59	333
Balance	-274	0	-159	-172	+158	+6	+265	+845

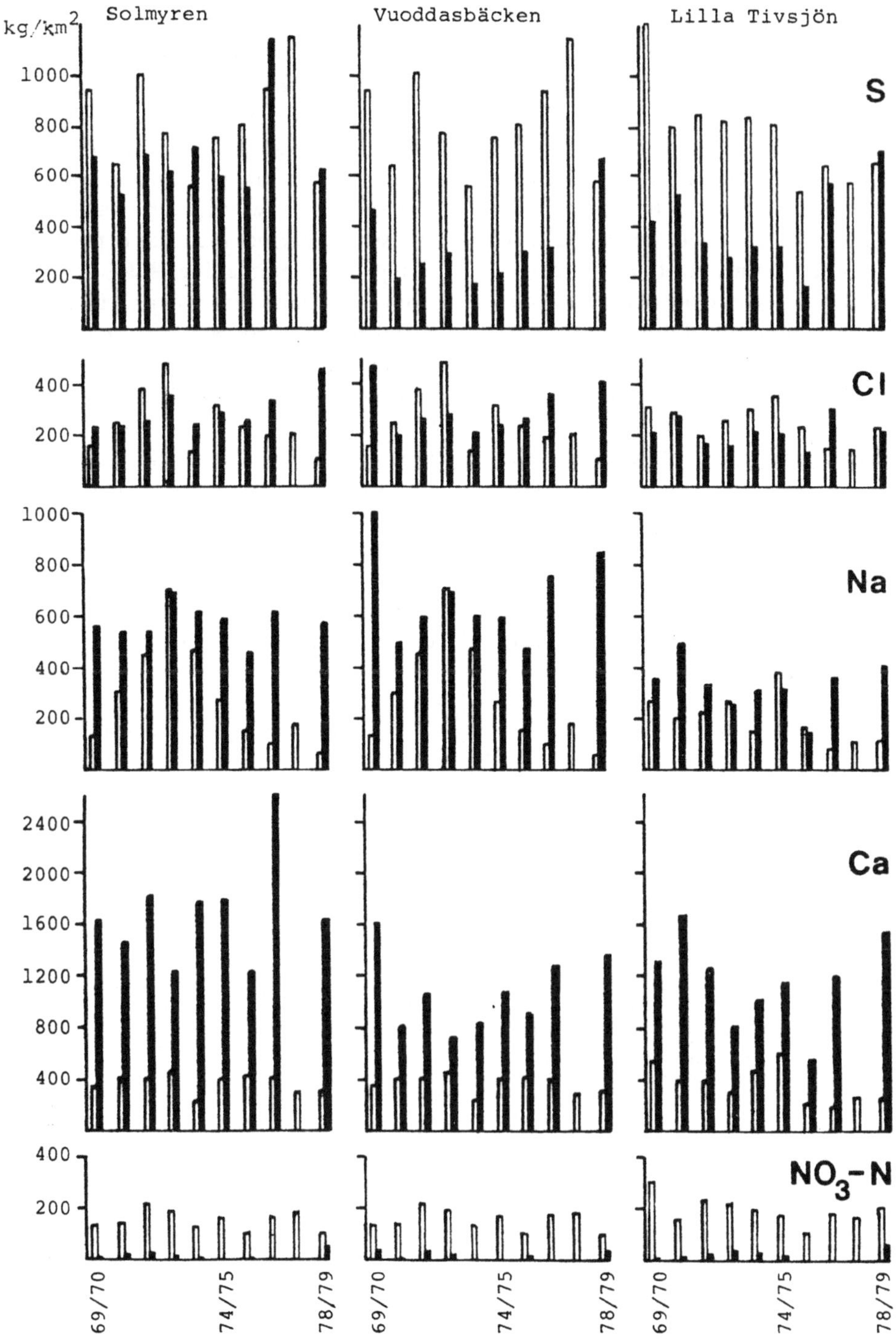

Fig. 2. Annual amounts of deposition from the atmosphere and discharge of five constituents in the three basins in kg · km ^{2}. Time unit is hydrological years.

course one can argue that the sulphur depositions have increased subatantially during the last decades because of an increase in fossil fuel combustion. Because of this the soil organic matter is in a process of readjustment to new 'equilibrium' levels accounting thus for the deficits. There is of course a considerable storage of sulphur in soil organic matter.

The other explanation is that part of the deposited sulphur is lost by release of volatile sulphur compounds, such as hydrogen sulphide and organic sulphur compounds like methyl sulphide and dimethyl sulphide. Very little seems to be known about these possibilities although the mechanism of formation of dimethyl sulphide from hydrogen sulphide is about the same as that forming dimethyl mercury. Considering the microbiological activity in soils loss of sulphur to the atmosphere from the soil is not unlikely.

The nitrogen compounds show deficits which are almost as large as the depositions. However, organic-bound nitrogen has not been determined but is certainly not greater than inorganic nitrogen runoff. On the other hand precipitation also contains organic nitrogen which will at least partly escape analytical procedures used and this certainly compensates the organic nitrogen in runoff. Hence, the deficits expressed as negative balances are certainly of the correct order of magnitude. Also, in this case there are two possibilities open to account for the deficits. One is an increase in soil organic nitrogen for the very same reason as for sulphur. The other possible explanation is denitrification processes in the soil which convert inorganic water-soluble compounds into nitrogen and nitrous oxide, N_2O that escape to the atmosphere. Another possible escape seems to be through nitric acid HNO_3 which can decompose and form volatile nitric oxide, NO. Again, it is possible that both processes – increase in soil organic nitrogen and denitrification – account for the deficits in the nitrogen balance.

The cations show in general a positive balance. The difference must then be accounted for by weathering of silicate minerals. Silicate minerals of interest in these hard rocks are mainly feldspars, biotite and hornblende. Of the feldspars the lime feldspar, $CaO \cdot Al_2O_3 \cdot 2\,SiO_2$ seems to be most easily weathered followed by the plagioclases – sodalime feldspars – soda feldspar, biotite, hornblende, all of which are reasonably easily discomposed by carbonic acid. The potash feldspars and muscovite are considered rather resistant towards weathering. Thus feldspars would supply sodium and calcium, and biotite would yield magnesium and potassium.

Temporal variations

The calculation of balance is, as mentioned in Table 2, based on mean values for a 10-year period. Within this period variations are large both between individual hydrological years and between

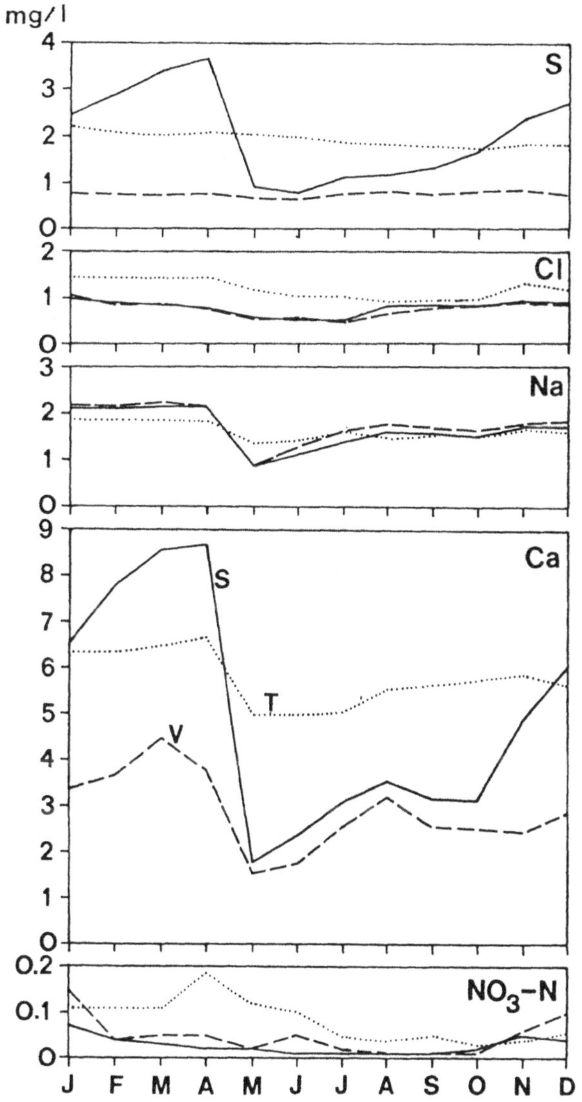

Fig. 3. Monthly means of concentration of five constituents in the three basins. V = Vuoddasbäcken, S = Solmysen, T = Lilla Tivsjön. Period 1969/70 1978/79.

monthly means of concentration of an individual constituent.

In order to illustrate the variation between different years the annual amounts of deposition and discharge is shown in Fig. 2 for S, Cl, Na, Ca and NO_3-N. Sulphur deposition by precipitation generally exceeds the discharge of sulphur except during the last years.

In Solmyren the discharged amounts of sulphur are closer to the deposited amounts than in other basins. Solmyren contains a large bog area.

The high calcium values in Solmyren in 1977 are due to the fact that a number of months that year showed very high values.

The monthly means of concentration of five of the constituents are shown in Fig. 3. For the Solmyren basin sulphur shows a strong seasonal variation similar to that of sodium and calcium. The concentration peaks of sulphur are also higher here than in the two other basins. Persson (1976) noted exceptionally high summer discharges in the Solmyren basin compared with other sub-basins within the Lappträsket basin. No satisfactory explanation for this has been given, but the high discharges during summer may be of importance when considering the high amounts of sulphur discharged.

The concentrations of calcium and sodium show a sharp drop in concentration during snow melt and in the case of calcium again Solmyren shows the widest range of variation during the year. Nitrogen compounds have very low concentrations for all the basins with minimum values occurring during late summer.

References

Andersson, U. M. & Eriksson, E., 1978. Hydrochemical investigations in three representative basins in Sweden. IHD Rep. 50, Stockholm.

Andersson-Calles, U. M. & Eriksson, E., 1979. Mass balance of dissolved inorganic substances in Sweden. Nord. Hydrol. 0: 99–114.

Anonymus, 1972. Hydrochemical analyses. A catalogue of procedures used within the IHD in the Nordic countries. Nord. IHD Rep. 3, Oslo.

Eriksson, E., 1952. Composition of atmospheric precipitation. Tellus 4.

Eriksson, E., 1955. Airborne salts and the chemical composition of river waters. Tellus 7.

Eriksson, E., (ed.) 1958. The chemical climate and saline soils in the arid zone. Arid zone Res. X. Reviews of research, pp. 147–180. UNESCO, Paris.

Eriksson, E., 1959. The yearly circulation of chloride and sulphur in nature; meteorological, geochemical and pedological implications. I. Tellus 11: 375–403.

Eriksson, E., 1960. The yearly circulation of chloride and sulphur in nature; meteorological, geochemical and pedological implications. II. Tellus 12: 63–109.

Eriksson, E., 1974. Transporthastigheten (flöden) ur statistisk synpunkt. In: Eriksson, E. & ?. Holtan (eds.). Hydrokemi, kemiska processer i vattnets kretslopp, pp. 83–88. Nord. IHD Rep. 7, Oslo.

Persson, M., 1971. Hydrologiska undersökningar i Lappträskets Representativa område. SMHI Nopr. 12: 000–000.

Persson, M., 1976. Hydrologiska undersökningar i Lappträskets Representativa område. Medd. VIII, SMHI, HB Rapp. 13. 000–000.

Waldenström, A., 1977: Slutrapport över hydrologiska undersökningar i Kassjöåns Representativa område. SMHI, HB Rapp. 29.

Dilution and acidification effects during the spring flood of four Swedish mountain brooks

Bruno Bjärnborg
The National Swedish Environment Protection Board, Water Research Laboratory, Box 1302, S-171 25 Solna, Sweden

Keywords: forest water ecosystems, water chemistry, snow chemistry, acidification

Abstract

The seasonal variation in water chemistry was followed during 1980 and 1981 in four mountain brooks in the southern part of Swedish Lapland. In the area investigated the soil is calcareous and the brook water is very well buffered during the major part of the year, with alkalinity varying between 0.4 and 1.0 milliequivalents per liter and with pH values about 7.5. These years the snow had a pH of approximately 5.2, which was considerably higher than has been reported from adjacent areas in the lower, coniferous region.

During snowmelt the water discharge increased drastically, and although the net transport of bicarbonate increased, alkalinity showed low values due to dilution with meltwater. pH decreased, but not further than to 6.3–6.5, far from values reported in 1979 (pH less than 5), apparently due to the comparatively clean snow. A slight deficit in alkalinity, as compared to the nonmarine calcium and magnesium content, points to an acidification impact on the area.

During maximum runoff some chemical variables, like aluminium, iron, nitrogen and phosphorus, behaved reversely to what might be expected during dilution and reached maxima in concentrations.

It is concluded that the extreme runoff characteristics of high mountain areas make brook water more sensitive to acid precipitation than might be expected when regarding only the calcareous properties of soil and bedrock.

Introduction

In 1979 the regional authorities in the county of Västerbotten made a survey of pH values in snow and brook water in the mountain areas and found low pH values in snow (mean pH 4.4) and in brook water during the spring flood (pH 4.9), in spite of normally well-buffered brook water. This was regarded as an indication that acidification had reached even the remote, low-polluted areas of Lapland. Acidification is also regarded as one possible cause for declining trout populations in the brooks (Fisk & Gydemo 1979).

In this study the water chemistry of a few selected brooks has been followed over a year cycle, as an attempt to verify the supposed acidification and to demonstrate the dilution mechanism during the spring flood.

The area

The sampling area is situated in southern Swedish Lapland, close to the border between Sweden and Norway (66°0′N, 15°0′E), in the upper parts of the drainage area of the river Umeälven. See map in Fig. 1.

The area is dominated by the mountains Artfjället (1 469 m above sea level, asl) and Norra Storfjället (1 767 m asl). The altitude of the valley of Umeälven is between 450 and 520 m asl. In Fig. 1 two altitudes are marked, 700 and 1 000 m asl,

Hydrobiologia 101, 19–26 (1983).

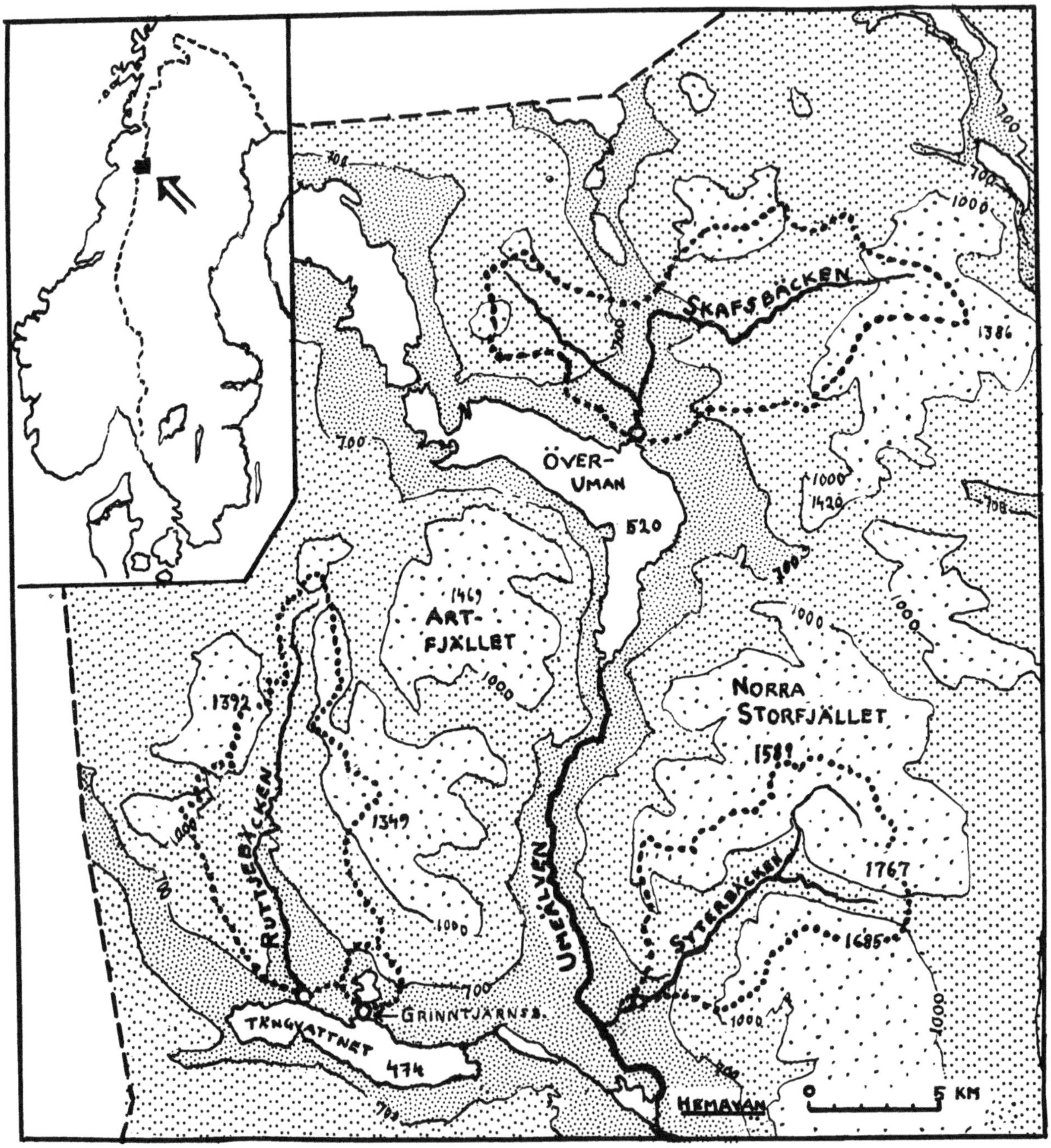

Fig. 1. Map of the investigated area.

.... = outlines of the catchment areas

⁘ = birchwood region

• = low-alpine vegetation zone

• = medium and high-alpine vegetation zone

approximately corresponding to the tree borderline and the upper limit of the low-alpine vegetation region, respectively. The catchment areas of the four brooks under investigation are also marked.

The brooks are Grinntjärnsbäcken (catchment area 3 km^2), Ruttjebäcken (60 km^2), Syterbäcken (62 km^2) and Skafsbäcken (77 km^2). The lengths of the latter three brooks are between 15 and 20 km. They represent varying degrees of steepness and altitudes, with Skafsbäcken as the least steep and Syterbäcken as the steepest. The lower areas of the Skafsbäcken basin have a considerable amount of swamps. Syterbäcken drains a glacier.

The small brook Grinntjärnsbäcken is in the outflow of a small lake which represents some 20% of the whole drainage area of the brook. It is mainly situated within the birch-wood region.

The bedrock in the area is dominated by calcareous phyllites and mica schists.

Materials and methods

Water samples were taken with polyethylene bottles near the outlet of the brooks, by Mr Evert Cederlund, who lives in the area and who chose the best sampling occasions. The sampling was concentrated to the period of the spring flood. In 1980 the emphasis was put on the initial part, whereas in 1981 the phase of declining discharge was sampled to a higher degree.

When possible, snow samples were taken in connection with water sampling, usually some 200 m away from the roads. Snow profile samples were taken with a 1 m long plexiglass tube and care was taken not to include snow in contact with soil. The snow profiles were transferred into plastic bags, allowed to melt, and the meltwater was transferred into plastic bottles. All equipment had been washed with distilled water. Samples were immediately sent to the laboratory in Solna by mail (which, unfortunately, in some cases took up to one week).

The chemical analyses were performed according to Swedish Standard Methods when possible (pH, conductivity, colour, Σ Ca + Mg, Cl, alkalinity, nitrogen and phosphorus compounds, Al, Ca, Fe and Mn, see References). Sulphate was determined by photometric titration (thorin), K and Na by flame photometry and trace metals (Cd, Cr, Cu, Ni, Pb and Zn) by graphite furnace atomic absorption techniques.

The Swedish Meteorological and Hydrological Institute has a water level gauge station at the outlet of Lake Tängvattnet. From this station data on the daily mean discharge were obtained, see Fig. 2A. The brook Ruttjebäcken contributes 30% of the runoff in the drainage area of this station and it probably follows a similar discharge pattern as the whole Tängvattnet area. This water discharge pattern has been supposed to be representative also for the other two large brooks. Snow cover records measured at the village of Hemavan near Syterbäcken are also noted in the figure. These data are sampled from the monthly publications from the Swedish Meteorological and Hydrological Institute.

Results and discussion

Snow chemistry

Neither in 1980 nor in 1981 did the snow samples show pH values that were as low as those for 1979, reported from Fisk & Gydemo (1979). According to Table 1, pH values just before snowmelt were between 5.2 and 5.3. The table also shows that the ion composition was dominated largely by sodium chloride, which indicates a strong marine influence. The distance to the North Atlantic from the center of the area is about 100 km.

The sulphate content was rather low (0.02 meq $\cdot$ l^{-1}), as compared with the values found at the regular precipitation chemistry stations in northern Sweden (Abisko, Racksund and Bredkälen). The last two stations, however, are situated in the lower, forested parts of northern Sweden. The Abisko station is probably the most comparable with the area studied, although situated in an other part of the mountain region. Sulphate concentrations at Abisko were generally lower than at the other two stations (0.02–0.03 and 0.03–0.10 meq $\cdot$ l^{-1}, respectively), and the chloride to sulphate ratio was higher. The precipitation at Abisko was on the average more acid during the winter of 1978–79 than during the winter of 1979–80 (pH approximately 4.8 and 5.3, respectively). In a regional survey of snow chemistry in March 1980 in the adjacent county of Norrbotten pH was generally about 4.5, but somewhat higher in the mountain areas, where the sulphate content was rather high (0.06 meq $\cdot$ l^{-1}). The authors of the report considered the origin of sulphate in the mountain region to be primarily dissolved salts and not sulphuric acid (Nauwerck *et al.* 1981). Together these data give some support to the observation that snow was indeed more acid in 1979 than in 1980 in the sampling area.

The ionic composition of the snow varied rather much during the winter of 1981, starting with low concentrations, dominated by sodium chloride. In February, after a short thaw period, snow samples were taken above the tree line at an altitude of 800–900 m asl, 2–5 km from the roads for compari-

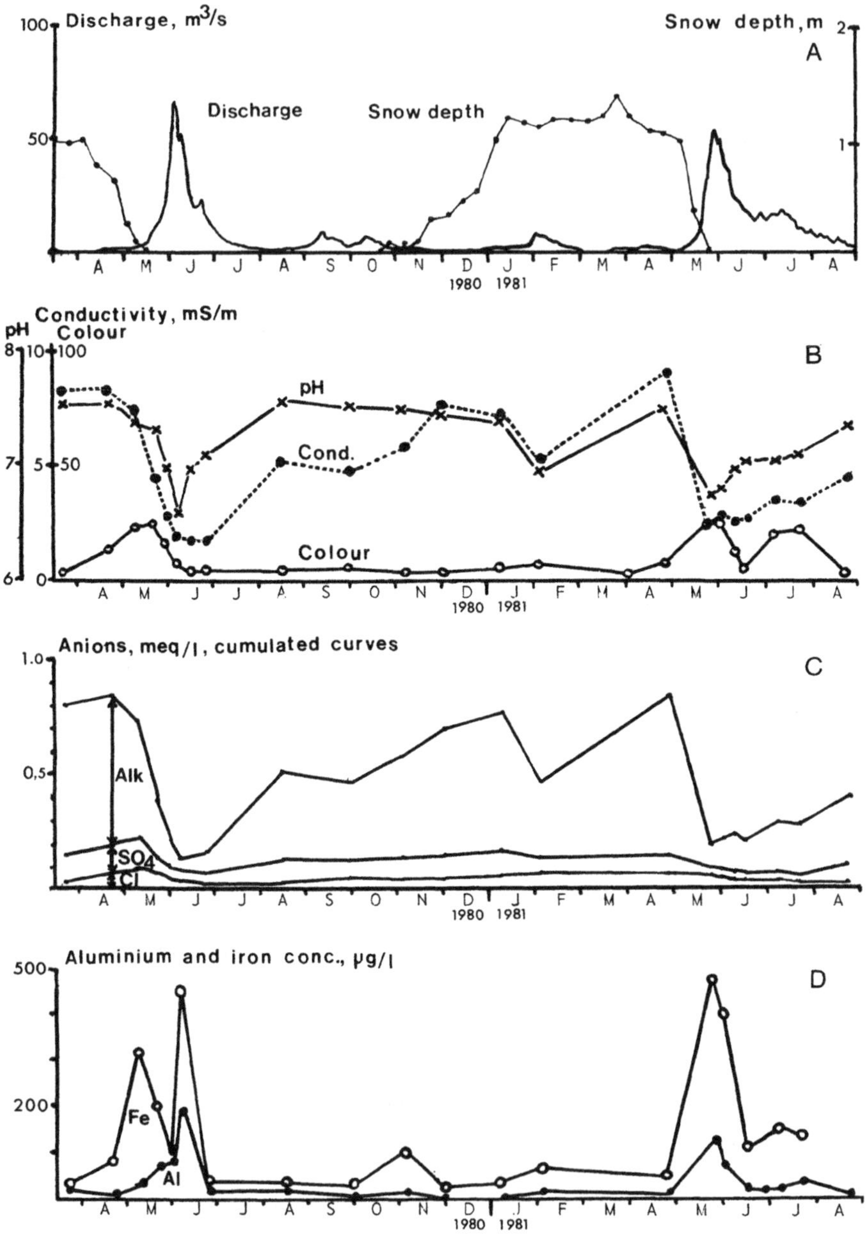

Fig. 2. A. The daily mean discharge at Tängvattnet and the snow cover at Hemavan.
B–D. Seasonal variations of chemical parameters in mountain brook water. Each point represents the mean value of measurements in the three brooks Ruttjebäcken, Skafsbäcken and Syterbäcken.
B. pH (x—x), conductivity in mS · m^{-1} (●· · ·●) and colour (○· · ·○).
C. Concentrations of anions in meq · l^{-1}. Curves are cumulated to show the total anion concentration.
D. Total concentrations of aluminium (●—●) and iron (○—○) in $\mu g \cdot l^{-1}$).

Table 1. Ion concentrations (meq · l^{-1}) in snow. Mean of three samples.

	pH	H^+	Na^+	K^+	Σ Ca + Mg	Ca^{2+}	NH_4^+	SO_4^{2-}	Cl^-	NO_3^-	alk.
1980-03-20 before snowmelt	5.2	0.006	0.05	<0.01	0.01	0.01	–	0.02	0.05	0.005	<0.01
1981-02-03 above the tree line	6.1	0.001	0.02	0.01	0.06	0.02	0.004	0.01	0.02	0.001	0.03
1981-04-27 before snowmelt	5.3	0.005	0.08	<0.01	0.03	0.01	0.005	0.02	0.10	0.006	<0.01

son. Thosê samples contained a remarkable amount of alkalinity, calcium and magnesium, which indicates that wind erosion can contribute basic dust, leading to neutralization of the snow acidity; pH was as high as 6.1 (Table 1).

The aluminium content of the snow was about 10 μg · l^{-1} but increased towards the end of the snow period to reach 30–60 μg · l^{-1}. The total phosphorus concentration behaved similarly, reaching levels of 10–30 μg · l^{-1}. The total nitrogen content, on the other hand, was rather constant within 200–400 μg · l^{-1}, with ammonium and nitrate representing only a minor fraction of the total nitrogen content. Of the heavy metals being analyzed, manganese and nickel seldom were detectable (<5 and <2 μg · l^{-1}, respectively) and other metals were close to their detection limits (Copper 0.5, Iron 20 and Zinc 10 μg · l^{-1}). Cadmium ranged between less than 0.05 and 0.2 μg · l^{-1} and lead between less than 0.5 and 3.3 μg · l^{-1}.

Brook water chemistry

The seasonal variations of the water chemistry of the brooks Ruttjebäcken, Skafsbäcken and Syterbäcken were quite similar to each other, and the data, therefore, are given as the mean values of the three brooks on each sampling occasion (Figs. 2B–D). Ruttjebäcken had the most well-buffered water, with a midwinter alkalinity exceeding 1.0 meq · l^{-1}, and the least marked pH drop during spring floods, while Syterbäcken had the least-buffered water, around 0.4 meq · l^{-1} in winter. Skafsbäcken had the greatest amplitudes in water colour.

First, it is worth noting that neither in 1980 nor in 1981 the pH drop during the spring flood was as deep as in 1979 (Fisk & Gydemo 1979). It is not probable, at least not in 1980, that a further heavy pH drop has been missed due to lack of sampling.

The streamwater chemistry during the spring flood generally followed the scheme proposed by Johannessen *et al.* (1980). According to this scheme, the first phase is characterized by an increase in the concentrations of calcium, magnesium and bicarbonate, due to the pressing out of 'old' water from the ground. This was observed in the samples from Syterbäcken and Skafsbäcken, with slightly increasing alkalinities, but in 1981 sampling was to sparse for the detection of this phase. In the second stage sulphate and chloride concentrations rose, due to ion separation in the snowpack and ion release in the early stage of the snowmelt. Then followed a third stage with dilution of all ionic species during maximum discharge. Eventually upon decreasing water flow, the weathering types of ions increased (calcium, magnesium and bicarbonate), while sodium, chloride and sulphate remained at lower values for a long time. It could, however, be noticed that the sulphate concentration never became lower than in snow. In 1981 the phase of recovery was very much drawn out.

Nitrate was present at low concentrations (maximum 0.01 meq · l^{-1}), showing variations similar to those of sulphate. The colour values fitted well to the scheme, but in 1981 high colour values also occurred at the point of maximum discharge and during the recovery period.

Other chemical components became more concentrated at maximum water flow, for example the total amounts of aluminium, iron, phosphorus and, to some extent, nitrogen. One reason might be that the very violent water flow whirls up and carries small soil particles. It is not known to what extent the compounds were dissolved, due to the total determination methods, but the nitrate and ammonium shares of the total-nitrogen peaks were small (2–12%), while the PO_4–P share of the total-phosphorus peaks was strongly varying (3–80%). The total-nitrogen concentrations increased from approximately 300 μg · l^{-1} to 600–1 000 μg · l^{-1} at flow peaks. The corresponding mean values for total-phosphorus were 4 and 75 $\mu g \cdot l^{-1}$, respectively.

Aluminium is of special interest, as it occurred in concentrations well above the levels that have negative impacts on trout at sufficiently low pH. However, pH never reached that low level. Considering the colour values, only some of the aluminium could be regarded as being bound to organic complexes. An attempt was made in 1981 to separate

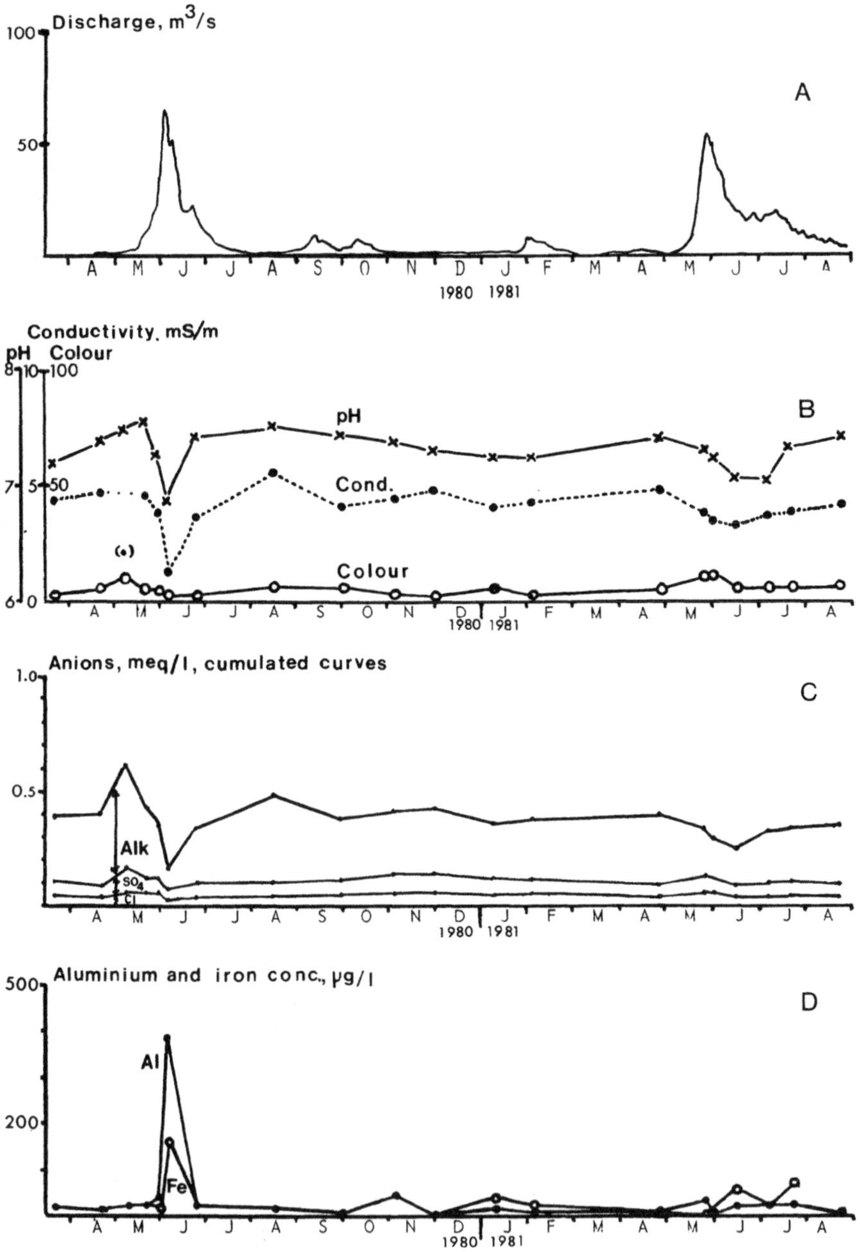

Fig. 3. A. The daily mean discharge at Tängvattnet.
B–D. Seasonal variations of chemical parameters in water of the brook Grinntjärnsbäcken.
B. pH (x—x), conductivity in $mS \cdot m^{-1}$ (●···●) and colour (O···O).
C. Concentrations of anions in $meq \cdot l^{-1}$. Curves are cumulated to show the total anion concentration.
D. Total concentration of aluminium (●—●) and iron (O—O) in $\mu g \cdot l^{-1}$).

the total aluminium into different fractions, either by an ion-exchange method (Driscoll 1980) or by direct analysis with the standard pyrochatecol-violet method used without the digestion step. The methods gave contradictory results: the ion-exchange results indicated low levels of polymeric or organically-bound aluminium, which in turn would indicate high levels of monomeric, inorganic aluminium, but the direct method gave very low values of monomeric (reactive) aluminium. One way to explain this result is that aluminium existed mainly as small particles, big enough to be retarded in the ion-exchange column. Even that, however, seems a little curious, as a coarse ion-exchange resin (20–50 mesh) was used to achieve high flow rates and low retention of suspended matter.

The iron concentration showed double peaks during the spring flood, with one peak coincident with the colour maximum, and one peak at maximum discharge (Fig. 2D). Manganese followed a pattern similar to that of iron, but at a ten-fold lower concentration level. The copper concentrations varied irregularly between <0.5 and $2.0\ \mu g \cdot l^{-1}$. The other heavy metals analyzed never exceeded the detection limits.

The small brook Grinntjärnsbäcken presented smaller seasonal variations in water chemistry (Fig. 3), apparently due to the detention and equalizing effect of the lake volume in the catchment area. As several days passed between sampling and analysis there might have been a change in some chemical species in the samples. Brook water at maximum discharge is probably not in equilibrium with suspended soil particles, but this stage will be approached during transport. Therefore, pH might have been lower if measured directly in the field (which was also the case on a few occasions), and concentrations of dissolved aluminium might have been higher.

Dilution mechanisms

There are two ways of regarding the dilution mechanism during the spring flood. One theory, a little generalized, presumes that there is a more or less constant base flow of ground and soil water, and that the meltwater is reaching the brooks mainly by surface runoff, leading to dilution of the base-flow water (discussed by Johannessen *et al.* 1980 and by Hårdén & Dahlskog 1980). The second theory is that the meltwater infiltrates via the ground to a much higher degree, which has been shown in southern Sweden by ^{18}O-analysis (Rodhe 1981).

However, the hydrological conditions in the high mountains can not immediately be compared with those in other parts of the country. The brooks investigated drain areas of most varying altitudes and the snow does not melt simultaneously over the whole area. Large areas consist of almost bare rocks. As can be seen from Fig. 4 the transport of bicarbonate in Ruttjebäcken reaches a maximum at maximum runoff. The snow at high altitudes has occasionally been shown to contain some bicarbonate (Table 1), but hardly at concentrations high enough to explain the increased transport of bicarbonate, which demands a contribution of bicarbonate from the ground and the soil. This indicates the dilution phase to involve other mechanisms than simply the dilution of a base flow with meltwater, impoverished in ions.

Acidification

A plot of alkalinity versus the sum of calcium and

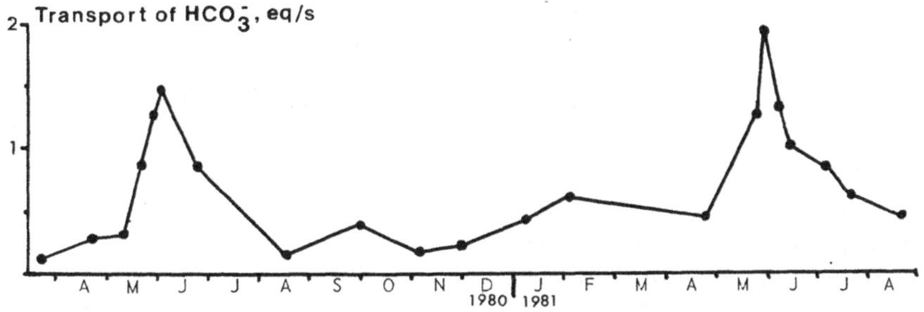

Fig. 4. Seasonal variations in the transport of bicarbonate in the brook Ruttjebäcken, equivalents per second.

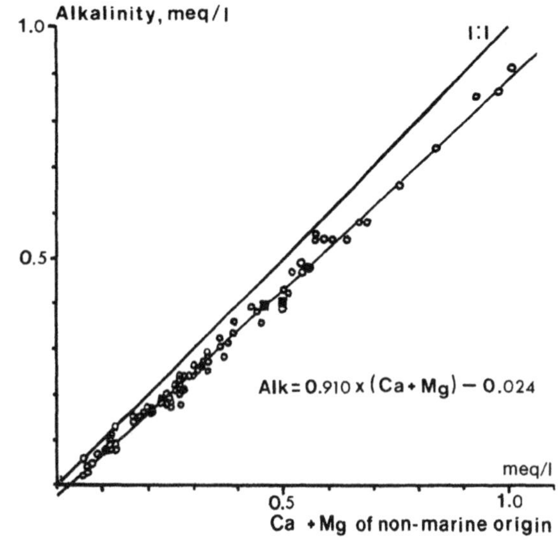

Fig. 5. The relationship between alkalinity (meq · l^{-1}) and the sum of concentrations of nonmarine calcium and magnesium (meq · l^{-1}) in all four brooks. Filled circle refers to double points and squares to triple points.

magnesium concentrations, of nonmarine origin, will show a deviation from the 1:1 line if the water has been exposed to acidification (Dickson 1980). When such a plot is done with data in this study (Fig. 5) it can be seen from the intercept that there is a lack of alkalinity of the magnitude of 0.02 to 0.03 meq · l^{-1}, which corresponds to the sulphate concentrations found in snow. An explanation of the deviating slope of the regession line might be the weathering of sulphates.

At maximum discharge the alkalinity reached values as low as 0.02 meq · l^{-1} in Syterbäcken 1980, in spite of moderately lowered pH values (5.2) in snow. The critical acidity of the snow can be roughly estimated, assuming the maximum infiltration capacity of the ground to be reached and no further contribution of alkalinity to occur. The buffering capacity would be enough for 0.02 meq · l^{-1} further load of hydronium ions via the meltwater. The snow in 1980 contained 0.006 meq · l^{-1} of hydronium ions, which gives a sum of 0.026 meq · l^{-1}. This corresponds to a pH in snow of 4.6, to be compared to the 1979 mean pH value of 4.4 found by Fisk & Gydemo (1979). Mountain brooks which are less buffered will, of course, be more sensitive to acid meltwater, and here pH values below 5.0 will become critical.

Conclusion

The extremely high water flows during the spring flood in the mountain areas, approximately 100 times higher than at 'base flow', lead to a drastic reduction of the buffer capacity of the brook water, making even drainage areas with calcareous soil sensitive to further acidification of the precipitation.

Acknowledgements

I would like to thank Evert Cederlund for help with sampling and William Dickson for inspiring discussions.

References

Dickson, W., 1978. Some effects of the acidification of Swedish lakes. Verh. int. Ver. Limnol. 20: 851–856.

Dickson, W., 1980. Properties of acidified waters. Norway 1980 SNSF-project. Proc. int. conf. ecol. impact acid precip. pp. 75–83.

Driscoll, C. T., 1980. Chemical characterization of some dilute acidified lakes and streams in the Adirondack region of New York State, Ph.D. thesis, Cornell Univ., 309 pp.

Fisk, E. & Gydemo, R., 1979. pH-mätning i snö och smältvatten i Västerbottens läns fjälltrakter våren 1979. Stencil from Fiskenämnden och Lantbruksnämnden i Västerbottens län, Umeå, Sweden, 17 pp. (in Swedish).

Hårdén, P-O. & Dahlskog, S., 1980. Acidity in the headwaters of the Lilla Lule River, Swedish Lapland, in 1979. Vannet i Norden 1: 3–25.

Johannessen, M., Skartveit, A. & Wright, R. F., 1980. Streamwater before, during and after snowmelt. Norway 1980 SNSF-projet. Proc. int. conf. ecol. impact acid precip. pp. 224–225.

Nauwerck, A., Lindqvist, L. & Groth, H., 1981. En nederbördskemisk snöinventering i Norrbottens län, mars 1980. Länsstyrelsen i Norrbottens län, Planeringsavdelningens rapportserie 1980, Luleå, Sweden, 11: 96 pp. (in Swedish).

Rodhe, A., 1981. Spring flood meltwater or groundwater? Nord. Hydrol. 12: 21–30.

Swedish Standard Methods (Svensk Standard), SS 02 81 series; No 21–24, 26, 27, 31–34, 36, 39, 41, 50, 52, 57 and 61. Available from: SIS, Box 3295, S-10366 Stockholm (in Swedish).

Trace metals in Swedish natural fresh waters

Hans Borg
The National Swedish Environment Protection Board, Water Research Laboratory, Box 1302, S-17125 Solna, Sweden

Keywords: forest water ecosystems, metals, fresh water, natural levels, acidification, humic substances, fractionation

Abstract

Water samples were collected from unpolluted waters in southern and northern Sweden to show the range of natural trace metal concentrations. The results do not indicate any significant differences in metal levels between the two study areas. Negative correlations with pH exist for total concentrations of Cd, Pb, Al, Mn and Zn. The dissolved (dialysable) metal fractions also increase with increasing acidity. The concentration of humic substances influences the distribution of Fe, Mn, Al, Pb, Cr, Co and As, as shown by positive correlations with water colour.

Increasing alkalinity has a limiting effect upon concentrations and annual fluctuations of Cd and Pb, and to some extent on Zn, in running waters.

Introduction

For most elements, the natural background levels of trace metals in freshwaters are not very well documented. The low concentrations present in water have traditionally caused analytical problems and there is always a great risk of contamination during sampling and analysis. Concentrations of some trace metals in river waters of Sweden and Norway, measured by neutron activation analysis, have been reported (Landström & Wenner 1965; Lenvik *et al.* 1978).

The purpose of this study is to show the range of natural concentrations of trace metals in Swedish freshwaters, and their dependence on fluctuations in water chemistry. To increase the accuracy of the measurements, parallel analysis with independent analytical techniques was performed when possible.

Sampling and analysis

Twenty lakes in the southwest (Halland-Småland) and 18 lakes in the middle and north of Sweden (Västmanland–Dalarna–Hälsingland–Jämtland) were studied. (The areas will be referred to in the text as 'S. Sweden' and 'N. Sweden', respectively.) The lakes are small (0.1–1.5 km^2) and situated far up in the drainage areas with no other lakes upstream. They are located in forested areas and are only influenced by long-range airborne pollutants. The majority of the lakes have low buffer capacity and a low total salt content and about half of them are acidified by acid precipitation (pH range in the lakes: 4.7–7.4). Three rivers in the north of Sweden were also included; Svartälven (Dalarna), Storån (Jämtland) and Petikån, a tributary to the river Skellefteälven (Västerbotten). The sampling stations were located upstream from any occurring effluents. Some chemical characteristics of the waters studied are shown in Table 1.

During March–April 1980, before snowmelt, water samples were collected in the lakes at the depth of 2.5 m (approx. 2 m below the ice cover) using a plexiglass Ruttner sampler. Samples were stored in FEP-teflon bottles and preserved with concentrated high-purity nitric acid (E. Merck, Suprapur,

Hydrobiologia 101, 27–34 (1983).

Table 1. Chemical characteristics of investigated waters.

Study areas	*n*	pH Mv ± SD	Conductivity $mS \cdot m^{-1}$	Alkalinity $meq \cdot l^{-1}$	Colour values $mg\ Pt \cdot l^{-1}$
Lakes					
Southern Sweden	20	5.58 ± 0.64	7.59 ± 1.70	<0.08 ± 0.12	95 ± 70
Northern Sweden	18	6.07 ± 0.60	3.76 ± 1.16	<0.28 ± 0.73	65 ± 55
Rivers					
(annual average 1980)					
Svartälven	7	5.83 ± 0.30	2.56 ± 0.22	0.017 ± 0.015	130 ± 45
Storån	5	6.86 ± 0.18	2.80 ± 1.07	0.088 ± 0.008	30 ± 10
Petikån	8	6.85 ± 0.29	4.93 ± 1.19	0.30 ± 0.07	70 ± 10

$2\ ml \cdot l^{-1}$). To avoid differing storage conditions at the different laboratories, the bottles were kept deep-frozen (-20 °C) prior to analysis. Sampling bottles had been cleaned by a procedure modified according to Patterson & Settle (1976) and described elsewhere (Borg 1981).

The dissolved metal fraction was determined by *in situ* dialysis, by a method similar to the one described by Beneš & Steinnes (1974). The dialysis cell consisted of a piece of dialysis tubing filled with water double distilled in quartz apparatus. The pore size of the membrane was 2.4 nm. The dialysis bags were immersed into the lakes at the depth of 2 m below the ice and taken up after 7 days. The contents were transferred to FEP-teflon bottles and preserved as described above.

The suspended fraction was collected by pressure filtration in the field. Nitrogen was used to produce pressure in a Niskin sampler and the particles were collected on polycarbonate membranes (Nuclepore, 0.4 μm pore size), which were dried and weighed prior to analysis.

The determination of metals was made by graphite furnace atomic absorption spectroscopy (AAS), hydride-flameless AAS (As only), differential pulse anodic stripping voltammetry (ASV) and neutron activation analysis (NAA), which was used both for water samples and for filters.

Results and discussion

Analytical comparison

The first step of this work was an intercalibration of 4–5 independent methods. The results showed a total interlaboratory standard deviation of 30–70% for cadmium and lead, and 15–55% for copper, with the highest figures for the samples with the lowest metal concentrations. Good agreement was generally obtained between AAS- and ASV-analyses of lead, cadmium and copper and between AAS- and NAA-analyses of iron, arsenic and aluminium. The results have been reported by Borg (1981). Some of the water samples in this study were analysed by both AAS and ASV. The results of the comparison are shown in Table 2. Highly significant relationships exist for lead, cadmium and copper even if there is a considerable scattering, especially in copper measurements. Some indications of systematic differences are also apparent in the deviations from 1.0 in the slope of the regression line. Lead determinations show the greatest deviation, where the ASV-technique seems to give slightly higher values than AAS. The natural background level of lead in water, as reported in the literature, has been steadily decreasing through the years, as a consequence of increasing sensitivity in the analytical techniques and increasing awareness of contamination problems. An average concentration as low as $0.015\ \mu g \cdot l^{-1}$ in freshwater has been reported (Hirao & Patterson 1974). Consequently, results from the AAS-measurements will be used in the evaluations of the lead levels below.

Table 2. Regression analysis of AAS (x) vs ASV (y).

Element	Corr. coeff. *r*	Slope	*y*-intcpt	*n*	Significance
Lead	0.8381	1.74	-0.0096	20	$p < 0.01$
Cadmium	0.7189	0.48	-0.0009	20	$p < 0.01$
Copper	0.6396	0.76	0.185	20	$p < 0.01$

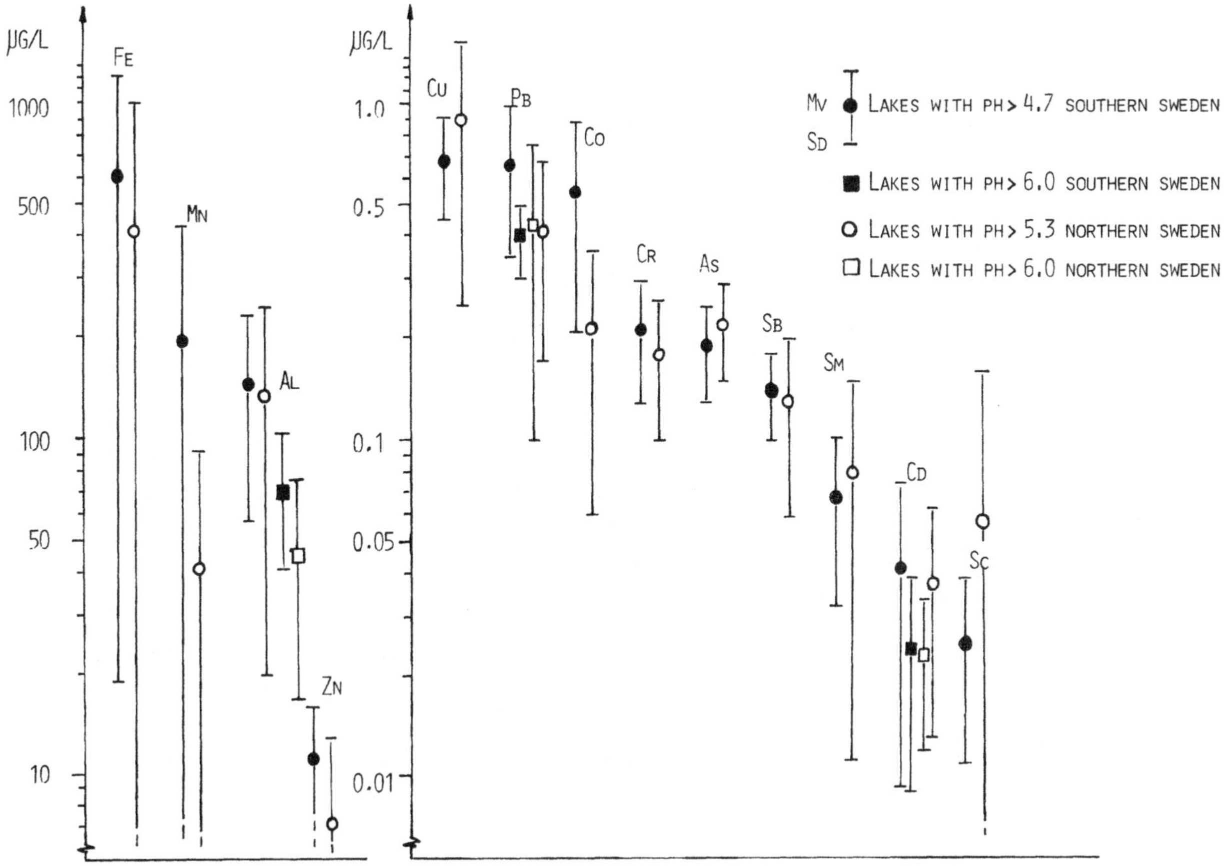

Fig. 1. Trace metal levels in Swedish woodland lakes, March–April 1980 ($\mu g \cdot l^{-1}$).

Distribution of total metal concentrations

Concentration ranges in southern and northern Sweden are shown in Fig. 1 and Table 3. In spite of the higher anthropogenic load on southern Sweden (as shown by Rühling & Skärby 1979) there were no appreciable differences in metal levels between the two study areas. The mean values for cobolt, zinc, lead and manganese, however, were higher in southern Sweden.

It should be pointed out here that the material includes some very acidified lakes, especially in the southern area. If only lakes with pH values above 6.0 were included the mean values for Al, Pb and Cd were lower. The Cd averages decreased from 0.038 and 0.042 to 0.023 and 0.024 $\mu g \cdot l^{-1}$ in northern and southern Sweden respectively. The total range of Cd concentrations in unacidified waters in this study was 0.004–0.05 $\mu g/l$ (Table 3 and Fig. 1).

Influence of pH and humic substances on total metal concentrations

Considering the chemical equilibria which limit the concentration of trace metals in fresh water, a decrease in pH would favour a prolonged retention time for some metals in the free water. That has also been demonstrated in lakes of different pH levels on the Swedish West Coast. Negative correlations were found between pH and Cd, Zn, Pb and Mn (Dickson 1980).

As no difference in metal concentrations can be observed between the northern and the southern study areas (see last section), the two sets of data were put together and checked for correlations with pH by linear regression analysis. As the pH value is the minus logarithm for the hydrogen ion concentration, the metal values were also transferred to logarithms.

Negative correlations with pH exist for Cd, Pb,

Table 3. Trace metal levels in Swedish woodland lakes.

Element	$\mu g \cdot l^{-1}$ Southern Sweden				Northern Sweden				Determination method
	Mv	± SD	Range	*n*	Mv	± SD	Range	*n*	
Fe	616	597	50–2 200	20	416	601	40–2 600	18	NAA
Fe (susp)	125	53	51–167	6	55	55	<4–184	9	NAA
Mn	195	238	8–553	16	42	49	2–149	17	NAA
Mn (susp)	5.8	10.4	0.4–27	6	1.5	1.2	0.2–4.4	10	NAA
Al	145	88	44–306	20	134	114	9–361	18	NAA
Al*	69	29	44–128	7	47	30	9–86	9	NAA
Al (susp)	12	6	3–19	6	14	16	3–51	10	NAA
Zn	11	5	4–25	15	7	6	<2–20	18	NAA
Cu	0.68	0.23	0.3–1.0	20	0.90	0.65	0.25–2.66	17	ASV
Pb	0.67	0.32	0.3–1.1	7	0.42	0.25	0.2–1.1	10	AAS
Pb*	0.40	0.10	0.3–0.5	3	0.43	0.33	0.2–1.1	6	AAS
Co	0.55	0.34	<0.2–1.2	16	0.21	0.15	0.1–0.7	18	NAA
Ni	<1.0	–	<0.9–1.2	6	<0.9	–	–	10	AAS
La	0.38	0.26	<0.1–1.0	16	0.39	0.31	<0.1–1.1	18	NAA
Cr	0.21	0.08	0.11–0.35	7	0.18	0.08	0.09–0.33	10	AAS
As	0.19	0.06	0.10–0.36	20	0.22	0.07	0.11–0.40	17	AAS
Se	<0.3	–	<0.1–0.4	16	<0.2	–	<0.1–1.0	18	NAA
Sb	0.14	0.04	0.08–0.19	16	0.13	0.07	0.06–0.17	18	NAA
Sm	0.07	0.03	0.03–0.14	16	0.08	0.07	0.02–0.26	18	NAA
Cd	0.042	0.033	0.005–0.12	19	0.038	0.025	0.004–0.09	16	ASV
Cd*	0.024	0.015	0.01–0.05	7	0.023	0.011	0.004–0.04	9	ASV
Sc	0.025	0.014	0.01–0.05	16	0.059	0.104	<0.005–0.46	18	NAA
Mo	<0.8	–	<0.1–2.4	16	<0.8	–	<0.1–1.5	17	NAA

Susp. = particulate fraction (⩾ 0.4 μm).
* = unacidified lakes only (pH ⩾ 6.0).

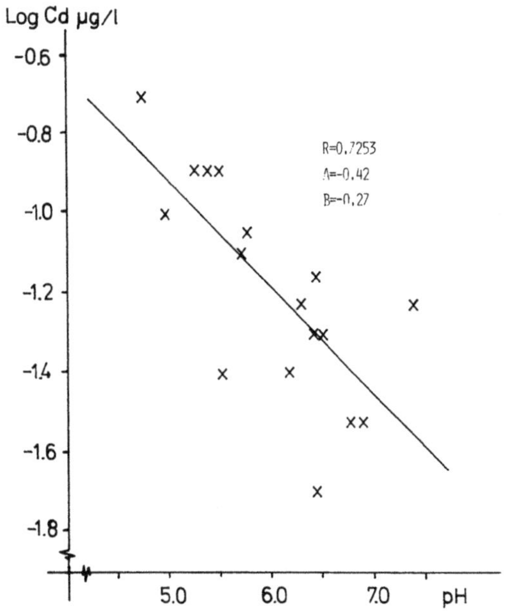

Fig. 2. pH vs log Cd in lake waters. R = correlation coefficient, A = *y*-intcpt, B = slope of the regression line.

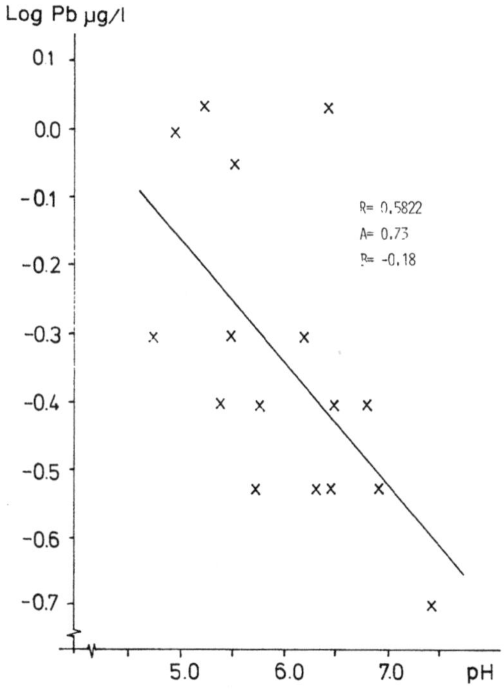

Fig. 3. pH vs log Pb in lake waters.

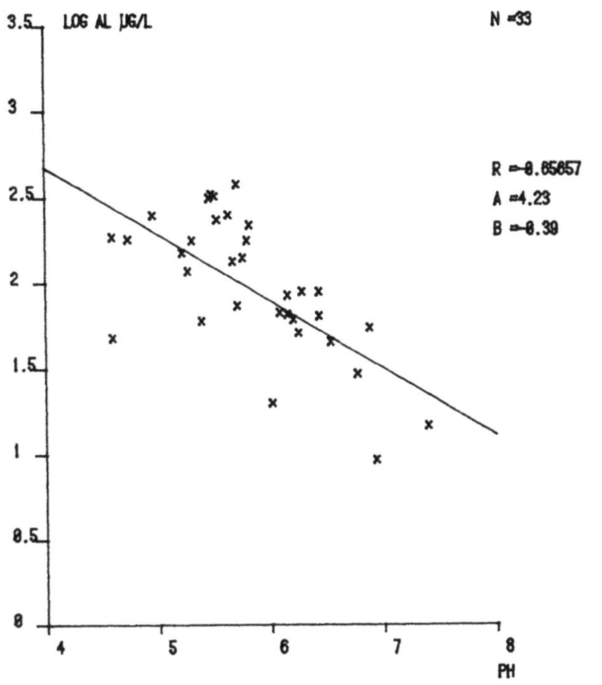

Fig. 4. pH vs log Al in lake waters.

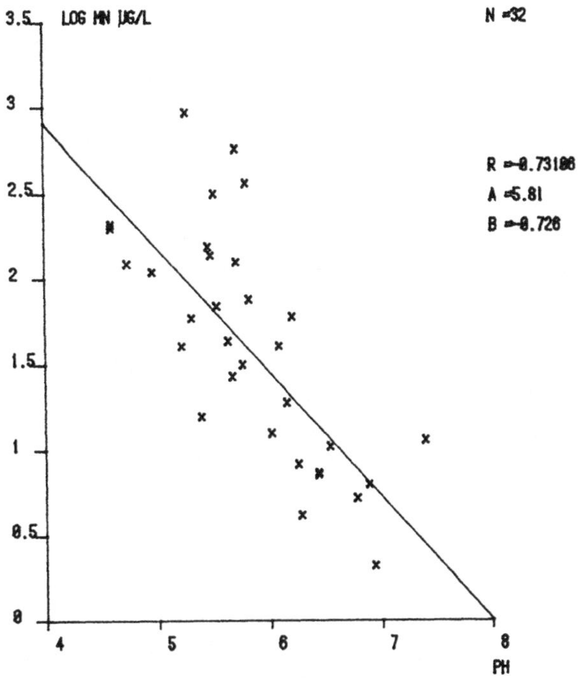

Fig. 6. pH vs log Mn in lake waters.

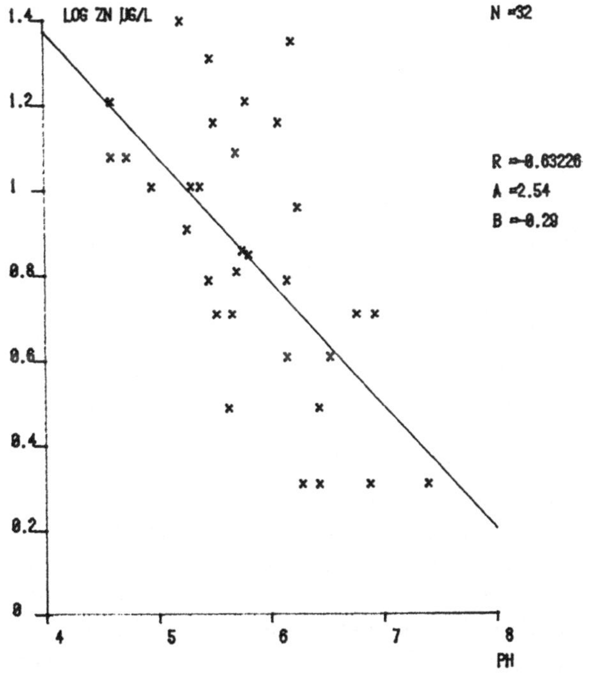

Fig. 5. pH vs log Zn in lake waters.

Al, Zn and Mn (Figs. 2–6). The relationships are statistically significant at the 99% confidence level (95% for Pt). The plot of Cd versus pH in Fig. 2 is based on the AAS-measurements. If ASV-measurements are used instead, the correlation coefficient is lower (0.4781 instead of 0.7253, $p < 0.01$ in both cases).

Another important factor which influences the trace metal levels in fresh water is the concentration of organics such as humic substances. Organic matter in water can be determined by TOC (total organic carbon) analysis but a simple estimate is offered by the widely used colour determination. To check the accuracy of the colour method it was compared

Table 4. Correlation between metal concentrations and colour in lake waters.

Metal	Correlation coefficient	*n*	Degree of significance
Fe	0.6424	31	$p < 0.01$
Mn	0.5665	32	$p < 0.01$
Al	0.4667	33	$p < 0.01$
Pb	0.5051	16	$p < 0.05$
Cr	0.6099	16	$p < 0.01$
Co	0.5126	32	$p < 0.01$
As	0.3628	37	$p < 0.05$

with TOC results on 66 samples from typical forest lakes. The correlation coefficient was 0.94, which is significant at the 99% confidence level. Consequently, colour values could be used as an approximation of the content of organic substances. Iron is typically associated with organics in freshwater and a significant positive correlation was found in this material between iron and colour values ($r = 0.64$) (Fig. 7), which was also demonstrated in similar lake waters in Västerbotten, northern Sweden, where the correlation coefficient was 0.97 (Borg 1978). Other elements which showed significant correlations with water colour were manganese, aluminium, lead, chromium, cobalt and arsenic. R values and degrees of significance are presented in Table 4.

Dissolved metal fraction

A few data on the dissolved metal fraction are available. The dialysis membranes used here had a pore size of 2.4 nm, which would permit free ions and small complexes to pass through, but not colloidal and particulate fractions. An increase of the dissolved fraction of most elements would be expected with increasing acidity of the lake waters. That was also the case for Pb, Al and possibly Cd (Fig. 8), in 'clear-water' lakes with colour values below 50. In lakes with higher concentrations of humic substances a greater part of Al and especially Pb seems to be associated with humic colloids which do not pass through the dialysis membrane. 10–20% of the Pb and 10–30% of the Al were dialysable in those lakes. About 5–25% of iron was dialysable, with slightly higher values in the more acidic lakes (Fig. 8). Beneš & Steinnes (1974) found about 9% Al and 20% Fe in the dialysable form in the Norwegian river Glomma. It should be pointed out here that there is of course a considerable uncertainty especially in the cadmium determinations because of the very low concentrations (0.003–0.05 $\mu g \cdot l^{-1}$).

Particulate metal fraction

The particulate fraction was collected in 16 of the lakes. Average values for Fe, Mn and Al are given in Table 3. The particulate metal fraction (in percent of total concentration) showed total ranges of 5–50 for Fe, 0.4–42 for Mn, 3–50 for Al and

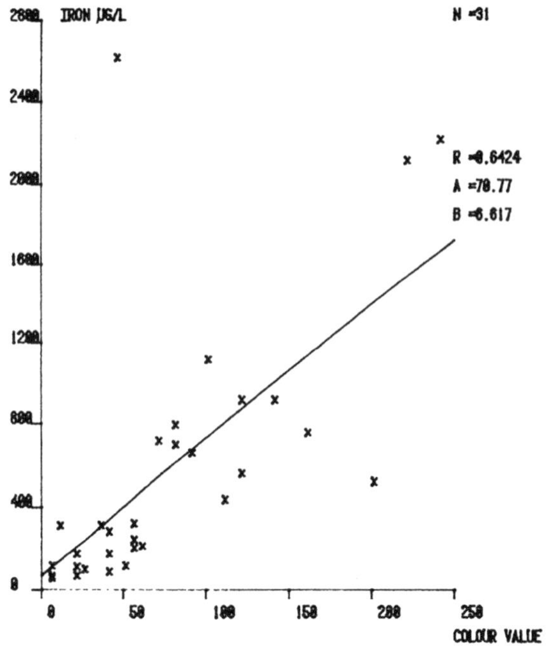

Fig. 7. Water colour vs iron concentration in lake waters.

1–10% for Zn. A significant positive correlation with pH was found for Zn ($r = 0.74$) and a weaker but similar trend for Mn and Al. The sample amounts were generally too small for most of the other elements to be detected with the NAA-method used for filter analyses.

Metal concentration in river waters

Some chemical characteristics for the rivers are given in Table 1. The river Svartälven has the lowest alkalinity, 0.017 meq $\cdot$ l^{-1} on an annual average, and the river Petikån the highest, 0.30. Svartälven has also the most acid water, the mean pH is 5.83 compared to 6.85 and 6.86 for the other two rivers. As might be expected, Svartälven has the highest average concentrations and the greatest fluctuations of zinc, cadmium and lead (Fig. 9). There is a tendency towards decreasing annual fluctuations of cadmium and lead with increasing alkalinity in the rivers. This is probably not caused simply by precipitation as metal carbonates in the river water, as the concentrations of lead, cadmium and carbonate ion at the actual pH and alkalinity levels are below the solubility products for the metal carbonates (log Ks = −13 to −14, Sillén & Martell 1964). The relationship might instead be due to the stronger binding of metals in

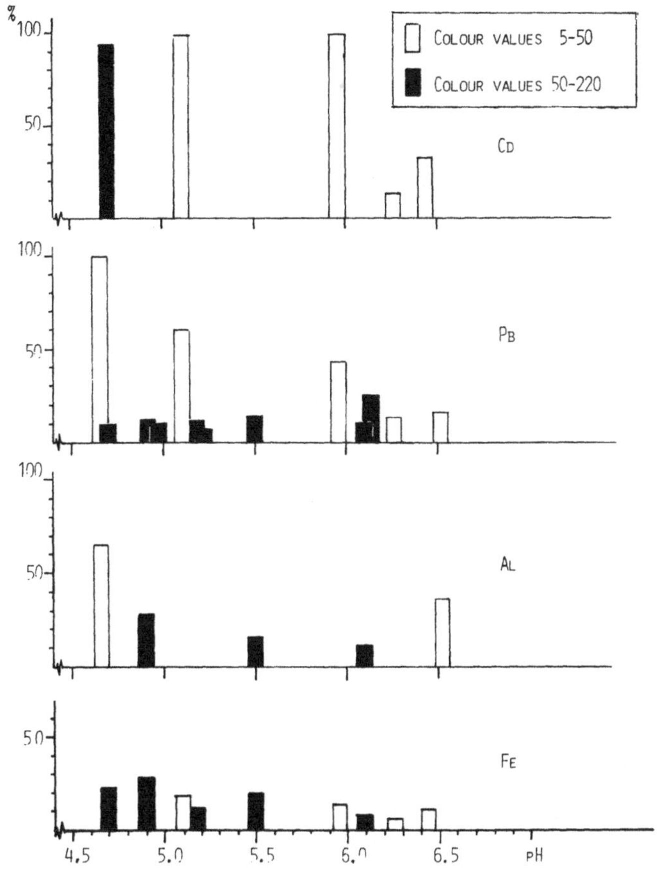

Fig. 8. Dissolved (dialysable) metal fraction in percent of total concentration in lake waters of different pH and colour (pore size 2.4 nm).

soils of more carbonate-rich watersheds, and to increased adsorption on mineral particles in water with a higher pH level, which causes increased sedimentation of metals. Besides the relatively low pH, the higher concentration of humic substances in the river Svartälven (Table 1) might favour an increased transport of lead, as lead is one of the elements that are associated to humus in freshwater.

Conclusions

No significant differences in the natural metal concentrations in northern and southern Sweden can be observed from these data. The main controlling factors for trace metals were shown to be pH, humic content and alkalinity. A negative pH correlation was found for both total metal concentrations (Cd, Pb, Al, Mn, Zn) and for dissolved (dialysable) metal fractions (Al, Pb). The humic content showed a positive correlation with the total concentrations of Fe, Pb, Cr, As, Al, Mn and Co, and was shown to limit the dialysable Pb fraction in humic rich waters.

Acknowledgements

NAA-measurements were performed by Jon-Per Rambaek and Eiliv Steinnes, Institute for Energy Technology, Kjeller, Norway, ASV-measurements by Ingemar Gustavsson, the National Swedish Environment Protection Board, Water Laboratory, Uppsala, and AAS-measurements by Mats Bengtsson, University of Lund, as well as the staff of the National Swedish Environment Protection Board, Water Laboratory, Solna.

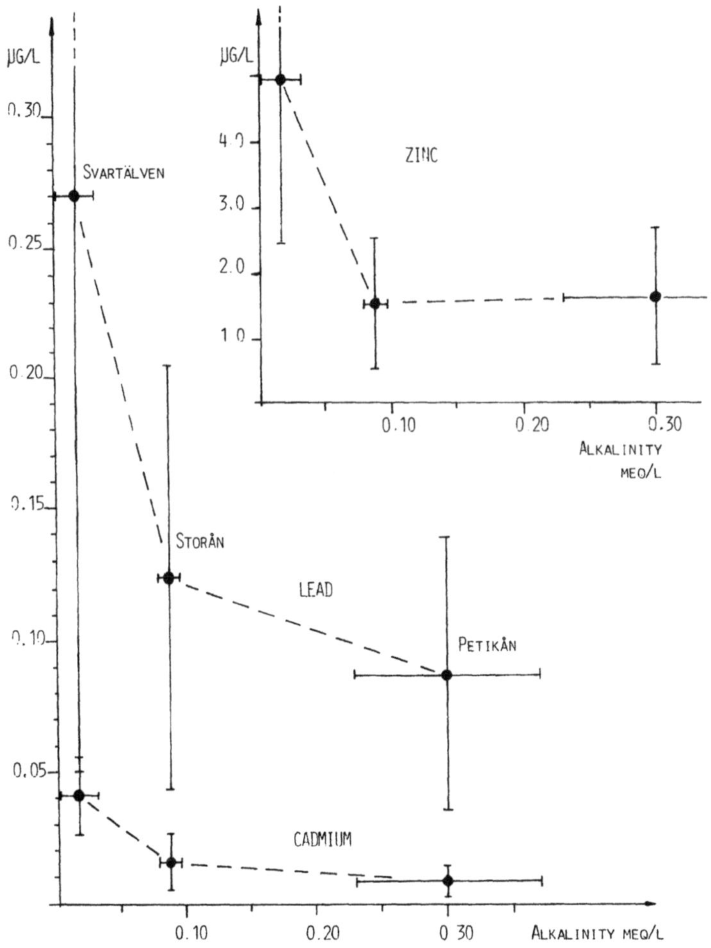

Fig. 9. Annual mean values and standard deviations of metal concentrations in three Swedish rivers with different alkalinity, 1980.

References

Beneš, P. & Steinnes, E., 1974. In situ dialysis for the determination of the state of trace elements in natural waters. Wat. Res. 8: 947–953.

Borg, H., 1978. Regional lake survey around the Rönnskär Smelter, northern Sweden. III. Impact of airborne metals on the metal content of lake waters. The National Swedish Environment Protection Board PM 1017, 24 pp, Solna, Sweden.

Borg, H., 1981. Trace metals in natural water. An analytical intercomparison. The National Swedish Environment Protection Board PM 1463, 28 pp, Solna, Sweden.

Dickson, W., 1980. Properties of acidified waters. In: Drabløs, D. & Tollan, A. (eds.). Ecological Impact of Acid Precipitation, pp. 75–83. SNSF-project, Ås, Norway.

Hirao, Y. & Patterson, C., 1974. Lead aerosol pollution in the High Sierra overrides natural mechanisms which exclude lead from a food chain. Science 184: 989–992.

Landström, O. & Wenner, C. G., 1965. Neutron-activation analysis of natural water applied to hydrogeology. Aktiebolaget Atomenergi, AE-204, 28 pp, Studsvik, Sweden.

Lenvik, K., Steinnes, E. & Pappas, A. C., 1978. Contents of some heavy metals in Norwegian rivers. Nord. Hydrol. 9: 197–206.

Patterson, C. C. & Settle, D. M., 1976. The reduction of orders of magnitude errors in lead analyses of biological materials and natural waters by evaluating and controlling the extent and sources of industrial lead contamination introduced during sample collecting, handling and analysis. In: LaFleur, P. (ed.). Accuracy in Trace Analysis, pp. 321–351. NBS' Special Publication 422, Wash. D.C.

Rühling, Å. & Skärby, L., 1979. National survey of regional heavy metal concentrations in moss. The National Swedish Environment Protection Board, PM 1191, 28 pp, Solna, Sweden.

Sillén, L. G. & Martell, A. E., 1964. Stability constants of metal-ion complexes pp. 140–141. The Chemical Society, London.

Selenium as a growth factor for plankton algae in laboratory experiments and in some Swedish lakes

Kåre Lindström
Institute of Limnology, Box 557, S-751 22 Uppsala 1, Sweden

Keywords: forest water ecosystems, selenium, algae, acidification

Abstract

The growth requirements of inorganic and organic selenium (Se) for the dinoflagellate *Peridinium cinctum* fa. *westii* and the diatom *Stephanodiscus hantzschii* var. *pusillus* and eight species of green algae are demonstrated.

A new mineral culture medium for *P. cinctum* is presented. Using *P. cinctum* as a test alga, bioassays were carried out on waters from a calciferous lake and from some acidified, fertilized, lime-enriched or humic lakes for determination of their contents of bioavailable forms of Se (bioactive Se). Some of the results were related to Tot-Se and selenite-Se, measured chemically, to pH and the occurrence of ambient phytoplankton.

In calcium-rich Lake Erken (pH $\sim$ 8) blooms and decline of algae coincided with decreases (from $>$70 to $\sim$20 ng Se $\cdot$ l^{-1}) and increases (up to $>$80 ng $\cdot$ l^{-1}) of bioactive Se, indicating uptake and release of inorganic or organic forms of Se. Acidified lakes (pH $<$ 5) generally demonstrated much lower concentrations of bioactive Se ($<$20 ng $\cdot$ l^{-1}) than neutral, fertilized, lime-enriched, slightly acidic or humic lakes ($>$20 ng $\cdot$ l^{-1}). The correlation between pH and bioactive Se was positive ($r \sim 0.92$; $n = 7$).

The ecological importance of Se is suggested for some common species of phytoplankton.

Introduction

Since Robinson (1933) found selenium (Se) toxic in grains and Schwarz & Foltz (1957) discovered it to be essential in mammal nutrition, considerable information about the existence and the consequences of this element in abiotic and biotic systems has been accumulating (Selenium and Tellurium Abstracts; Zingaro & Cooper 1974; Gamboa Lewis 1976; Swaine 1979).

Se is recognized and characterized as an essential trace element for animals and man, where it is involved in the function of various enzyme systems and biochemical processes, e.g. blood cells, muscle tissues and vital internal and reproductional organs (Vokal-Borek 1979; Zingaro & Cooper 1974). In the literature, however, e.g. agricultural science (Gissel-Nielsen 1977), plant physiology and microbiology (Shrift 1973) the discussion continues regarding its roles in nutrient metabolism and as a growth factor for plants (Committee on Medical and Biological Effects of Environmental Pollutants 1976).

The ability of Se to adapt various valences (+VI, +IV, +II, O, -II) makes it very mobile and bioavailable. For instance, Se-accumulator plants concentrate selenite and selenate in high quantities by incorporating and transforming them into organics (Se^{-II}), which are easily volatilized by methylation (Lewis *et al.* 1974) or released during decay. Many amino acids where Se is substituted for S have been identified in these investigated plants including the non-accumulators (Shrift 1973; Moxon & Olson 1974). Duckweeds, e.g. *Spirodela oligorrhiza,* accumulate and convert ionic Se^{+IV} or Se^{+VI}, which can be traced as it is incorporated into

Hydrobiologia 101, 35–48 (1983).

various proteins, such as Se-methionine, Se-cystine and Se-cysteine (Butler & Peterson 1967).

A few studies are published concerning uptake of Se in single-celled fresh water algae, e.g. *Scenedesmus* (Sandholm *et al.* 1973) and the marine *Tetraselmis* and *Dunaliella* (Wrench 1978). However, in these investigations there is only one report of growth-promoting effects on cellular proliferation of certain marine chrysomonads (Pintner & Provasoli 1968). Nor has any unambiguous response been documented with respect to Se as a growth factor for multicellular plants (Shrift 1973; Broyer *et al.* 1966; Committee on Medical and Biological... 1976) except for a recent study on the marine *Fucus spiralis* which showed significantly improved growth when minute concentrations of Se were added to the culture medium (L. Fries, pers. comm.).

Findings in microbiology point to Se as an important factor for the function of several bacterial enzymes such as formate dehydrogenase in *E. coli* (Pinsent 1954; Andreesen & Ljungdal 1973, 1974), glycine reductase (Stadtman 1974), nicotinic acid hydroxylase (Imhoff & Andreesen 1979), and xanthine dehydrogenase (Wagner & Andreesen 1977) in certain *Clostridia.*

Current knowledge of Se in aquatic systems is sparse and heterogenous, probably due to the difficulties in obtaining rapid and accurate measurements of the element at low concentrations (<10 ng $\cdot$ l^{-1}).

However, development of purified chemicals, modified and improved or new chemical analytical techniques (M. Remberger, pers. comm.) in combination with gas chromatography (Measures & Burton 1980) as well as biological methods for the determination of the bioavailable forms of Se (Lindström 1980) offer new tools for investigating the structure and metabolic role of Se in aquatic systems.

In limnology and oceanography data have been reported concerning concentration levels of the element (Traversy *et al.* 1975; Measures & Burton 1978; Lindström 1980), various stages of oxidation (Measures & Burton 1980) and transports, cycling and bioaccumulation (Rudd *et al.* 1980), but its ecological relevance is still unclear and unexplored.

The discovery of Se as an essential element for growth of the dinoflagellate *Peridinium cinctum* fa. *westii* (Lemm.) Lef. (Lindström & Rodhe 1978) has motivated further research regarding the general importance of this nutrient for algal growth and its occurrence in freshwaters. A temporal and spatial variation of Se was demonstrated in mesotrophic Lake Erken (eastern Uppland, Sweden) where phytoplankton growth was correlated with a decrease in the bioactive or the biologically available forms of Se as Se^{+IV}, Se^{+VI} and organic Se^{-II} (Lindström 1980). Increased Se concentrations occurred during periods of high precipitation, spring thaw, and during the decline of the algal population (op. cit.).

The aim of this study is to determine the growth responses of the dinoflagellate *P. cinctum* fa. *westii,* the diatom *Stephanodiscus hantzschii* var. *pusillus* Grun. and eight genera of green algae to inorganic and/or organic forms of Se.

The ecological significance of bioactive Se regarding phytoplankton growth in naturally calcium-rich Lake Erken during spring 1979 and the effect of pH on the bioavailability of Se in some calcium-rich, lime-enriched, fertilized, humic or acidified lakes in southern Sweden is also demonstrated.

Material and methods

Water samples were either filtered through glass fiber filters (GF/C) and stored at 4 °C for a short time or preserved at -20 °C for subsequent analysis (acidified and fertilized test water).

Bioactive selenium was determined by employing the *Peridinium*-bioassay technique according to Lindström (1980).

Inorganic Se was determined chemically based on Watkinson (1966) and Parker & Harvey (1962) but modified by Remberger, M. (pers. comm.). DAN (2,3-diaminonaphtalene)-chelated Se was extracted with n-hexane from 50 ml water samples and measured using fluorometry.

Counting of phytoplankton was performed by P. Blomqvist, using Utermöhl's technique and an inverted microscope. Accessory data were acquired from the Erken-limnological project (1976–1981; Swedish NFR No. B 2051-100).

Duplicate or up to five parallel series of batch cultures in Erlenmeyer flasks (100 ml) or test tubes (10 ml) were inoculated (0.1–0.5 ml) with low cell numbers (20–50 ind. ml^{-1}) of the axenic test alga *P. cinctum* fa. *westii.* The axenic conditions were regu-

Table 1. Lindström medium No. 16 (L 16) for *Peridinium cinctum* fa. *westii.*

Compound	mmol · l^{-1}	mg · l^{-1}	Element	mg · l^{-1}
Macronutrients				
$NaHCO_3$	1.0	84.00	C	12.01
			Na	23.00 } 26.45
$NaNO_3$	$1.5 \cdot 10^{-1}$	12.75	Na	3.45
			N	2.10
$CaCl_2 \cdot 2H_2O$	$1.0 \cdot 10^{-1}$	14.70	Ca	4.01
			Cl	7.09
$MgSO_4 \cdot 7H_2O$	$5.0 \cdot 10^{-2}$	12.32	Mg	1.22
			S	1.60
K_2HPO_4	$1.0 \cdot 10^{-2}$	1.74	P	0.31
			K	0.78
$Na_2SiO_3 \cdot 9H_2O$	$4.0 \cdot 10^{-3}$	1.14	Si	0.11
Micronutrients				
$FeCl_3 \cdot 6H_2O$	$4.0 \cdot 10^{-4}$	0.108	Fe	0.0223
$Na_2MoO_4 \cdot 2H_2O$	$2.0 \cdot 10^{-4}$	0.048	Mo	0.0192
$ZnCl_2$	$2.0 \cdot 10^{-4}$	0.027	Zn	0.0131
$MnCl_2 \cdot 4H_2O$	$5.0 \cdot 10^{-5}$	0.010	Mn	0.0028
$CoCl_2 \cdot 6H_2O$	$4.0 \cdot 10^{-5}$	0.0096	Co	0.0024
$Na_2SeO_3 \cdot 5H_2O$	$7.0 \cdot 10^{-7}$	0.000184	Se	0.000055
EDTA	$1.5 \cdot 10^{-3}$	0.558		
Multiples of L 16	1×	2×	4×	6×
Conductivity: $\mu S \cdot cm^{-1}$	140	280	560	770
Alkalinity: meq · l^{-1}	1.0	2.0	4.0	6.0
pH	8.1	8.4	8.7	8.9

larly checked on nutrient-enriched agar substrates (Oxoid, bactopeptone).

Various multiples of the L-medium No. 16 (L 16, Table 1) – a development of medium L 12 (Lindström & Rodhe 1978) – for *P. cinctum* fa. *westii* were used for all the culture experiments. The nutrient concentration levels in this medium are based on maximum growth rates of this alga (Lindström, in preparation).

Monocultures of bacteria-free *P. cinctum,* isolated from Lake Kinneret by A. Panders, and 6 out of 8 representants of non-axenic selenium-depleted green algae, isolated from Lake Erken by K. Lindström, were incubated at 12 h/d fluorescent light, 21.5 °C and 70–100 $\mu E \cdot m^{-2} \cdot s^{-1}$.

Non-axenic monocultures of the diatom *S. hantzschii* var. *pusillus* isolated from Lake Erken by C. Eriksson, were incubated at 15 °C and 12 h/d of fluorescent light (50 and 90 $\mu E \cdot m^{-2} \cdot s^{-1}$).

The cell growth was determined by measurements of relative *in vivo* fluorescence (F) of chl *a* at 683 nm (FM 3, Halldal), where F = 1 is ~1 μg chl *a* · l^{-1}. The dinoflagellate cultures were measured once a week, and the cultures of diatoms and the green algae every two to three days.

The production of cells is expressed as doublings day^{-1}:

$$K_F = \frac{{}^2\log F_t - {}^2\log F_0}{t},$$

where F = *in vivo* units of fluorescence and t = time in days.

The high correlation coefficient ($r > 0.97$), obtained for the cell numbers of *P. cinctum* (Lindström 1980) and *S. hantzschii* var. *pusillus* (Eriksson 1981) in relation to the numbers of fluorescence units indicates that K_F describes the division rate of the algae.

Results

Algal requirement of Se

An indispensable growth requirement of selenite ($Na_2SeO_3 \cdot 5H_2O$) has been shown for *P. cinctum* fa. *westii* (Lindström & Rodhe 1978; Lindström 1980). However, selenate added as $Na_2SeO_4 \cdot 10H_2O$ (instead of selenite as $Na_2SeO_3 \cdot 5H_2O$) to L 16 (Table 1), produced different growth responses of this alga, compared to additions of selenite as SeO_2 in the same medium (Fig. 1). During the growth phase of approximately maximum growth rate (days 0–26) the doublings day^{-1} were also rapid at only 5 ng selenite-Se, although $\sim k_{max}$ occurred within a range of 20 to 50 ng $Se \cdot l^{-1}$. A prolonged lag phase or adaptation period occurred in the selenate-enriched medium, which produced lower biomasses and growth rates within the first 26 days of algal growth, compared to the enrichments with selenite in the culture media. After this growth period highest growth rates were obtained with ad-

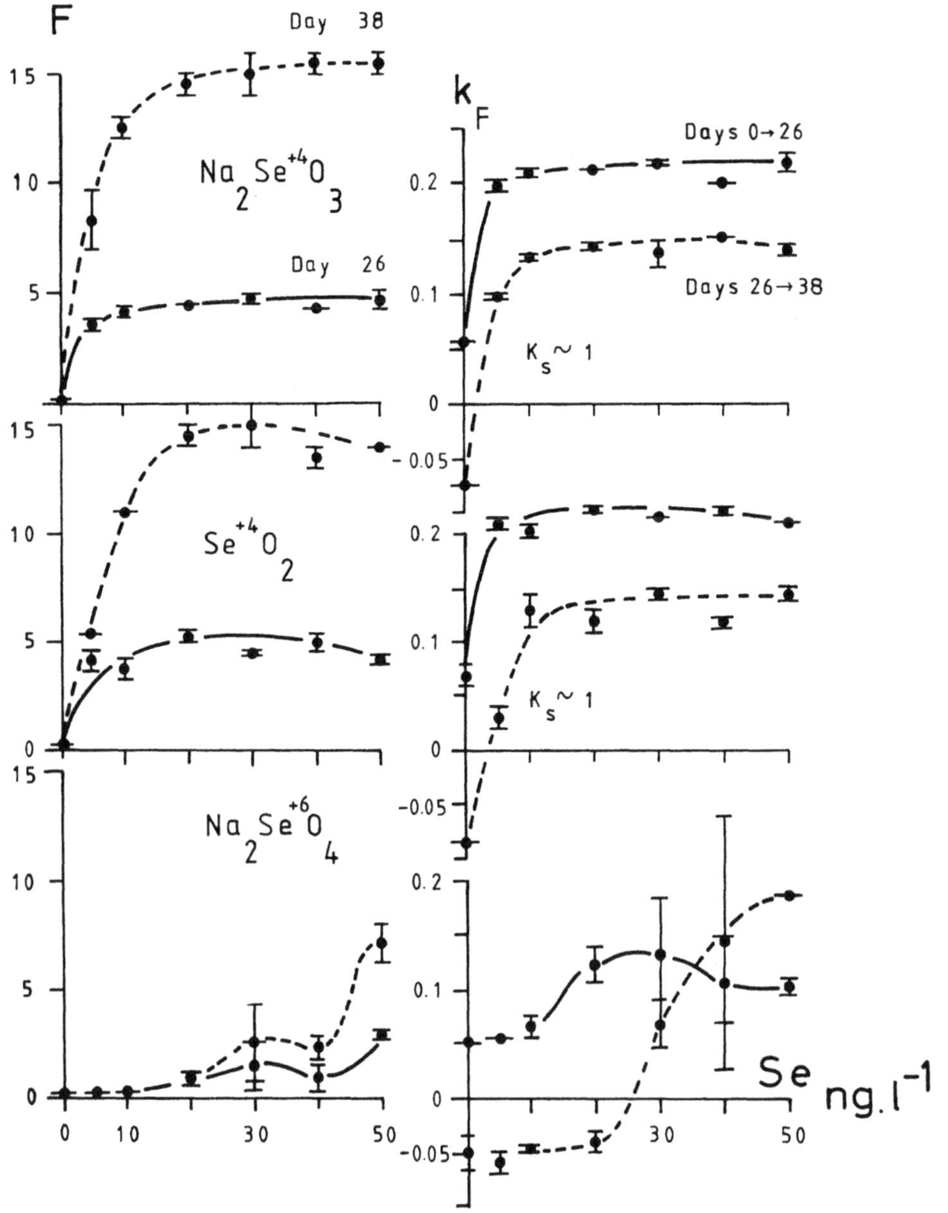

Fig. 1. Biomass development (F) and doublings $\cdot$ day^{-1} (K_F) of *Peridinium cinctum* fa. *westii* in relation to various concentrations of selenite-Se and selenate-Se in medium 2 × L 16 at 21.5 °C and ~90 $\mu E \cdot m^{-2} \cdot s^{-1}$. Two parallel experiments.

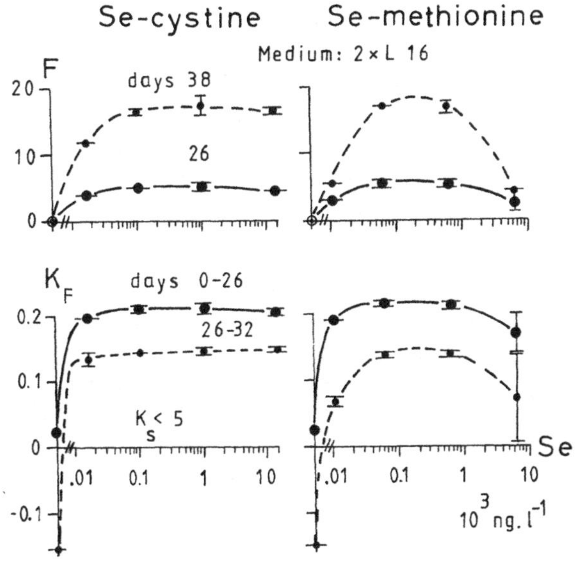

Fig. 2. Biomass development (F) and doublings · day^{-1} (K_F) of *Peridinium cinctum* fa. *westii* in relation to various concentrations of Se-cystine and Se-methionine in medium 2 × L 16 at 21.5 °C and ~90 μE · m^{-2} · s^{-1}. Two parallel experiments.

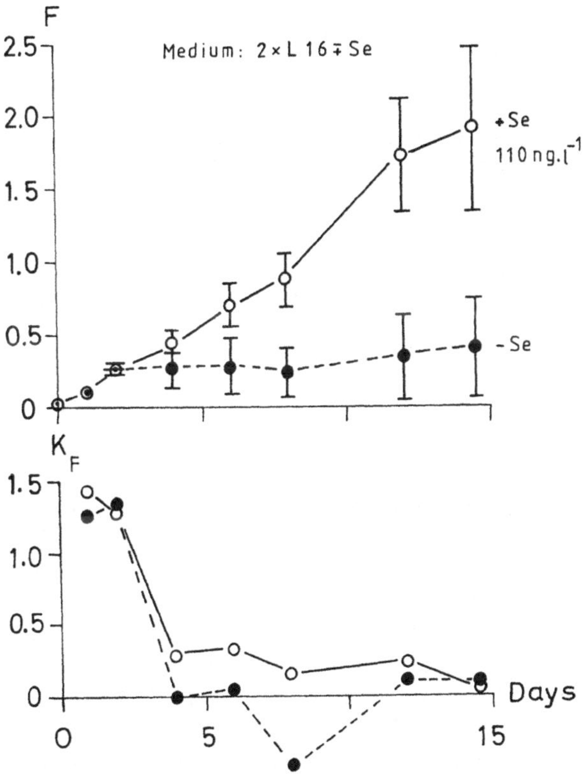

Fig. 3. Biomass development (F) and doublings · day^{-1} (K_F) of Se-depleted *Stephanodiscus hantzschii* var. *pusillus* grown in medium 2 × L 16 (110 ng selenite-Se · l^{-1}), and in the same medium without addition of Se. Five parallel experiments.

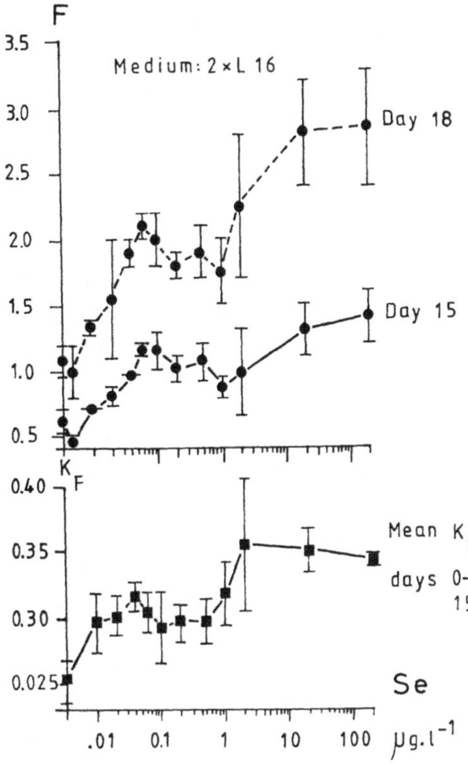

Fig. 4. Biomass development (F) and doublings · day^{-1} (K_F) of *Stephanodiscus hantzschii* var. *pusillus* grown in medium 2 × L 16 enriched with various concentrations of selenite-Se (ng · l^{-1}). Two parallel experiments.

ditions of 50 ng · l^{-1} selenate-Se. However, with this addition the biomasses later (>38 days) normally reached as high levels as in the selenite-Se-enriched cultures. Growth responses in several experiments indicate a more rapid uptake and/or more efficient mechanisms of this alga in utilizing selenite for growth, compared to selenate.

Table 2. Growth responses of nutrient-saturated and exponentially growing *Stephanodiscus hantzschii* var. *pusillus* to inorganic and organic Se in medium 2 × L 16. F_{max} = maximum biomass and $\bar{K}_F$ = mean doublings · day^{-1}. Two parallel experiments.

Medium 2 × L 16 ± Se	F_{max}	Day	$\bar{K}_F$	Growth period days
Se				
0 ng · l^{-1}	2.6–3.4	23	0.270–0.287	23
Se-cystine				
550 ng · l^{-1}	3.34–4.46	17	0.387–0.411	17
Se-methionine				
550 ng · l^{-1}	4.1–6.8	23	0.200–0.331	23

Organic Se as Se-cystine or Se-methionine was as good a nutrient source as inorganic Se for growth of *P. cinctum* (Fig. 2). Both biomasses and growth rates were highest, as with selenite, at additions of ~50 ng Se · l^{-1} ($K_s < 5$ ng · l^{-1}) of the organic Se compounds. These results indicate the importance of an organic fraction of the bioactive Se for algal growth.

A growth response to selenium was also obtained for the diatom *S. hantzschii* var. *pusillus* (Fig. 3), cultured in medium 2 × L 16 (Table 1). The inoculum algae were nutrient- and selenium-depleted. After 11 days of algal growth, the biomasses were up to 50 times higher with addition of 110 ng · l^{-1} selenite-Se in the culture medium, compared to no addition of this element. The growth rates were also much lowered without Se in the culture medium and the mean growth rate quote, K_F (+Se)/K_F (−Se), calculated for 14.5 days was 1.8. In medium 2 × L 16 (Table 1), high concentrations of selenite-Se (>0.05 μg · l^{-1}) further enhanced the growth of this diatom (Fig. 4). Optimum Se concentrations regarding biomass development occurred at ~0.05–200 ng Se · l^{-1} (days 15 and 18) and mean growth rate was highest at >2 μg Se · l^{-1} (days 0–15, 15–18).

The diatom is also capable of utilizing organic Se (Table 2). Compared to the cultures with no addition of Se, the addition of Se-methionine approximately doubled the maximum algal biomass (F_{max}), and addition of Se-cystine increased the mean growth rate to reach F_{max}. Cultures with no addition of Se also reached high biomasses due to the use of nutrient- and Se-saturated cells in the inoculum. A production of vitamins from ambient bacteria cannot be excluded as a stimulatory agent responsible for this divergence (see below).

The growth stimulatory properties of Se on *S. hantzschii* var. *pusillus* appear to be one factor for regulating the development of this diatom in natural waters.

Vitamin B12 also appears to be an important factor for the growth of this alga (Fig. 5). Various additions of thiamine, biotine and B12 to some cultures of the algal experiment presented in Fig. 3 showed, that of these vitamins, only B12 enhanced the growth of the alga (Fig. 5). A trace metal mixture containing 13 various elements did not effect the diatom growth. Additions of 100 μg B12 · l^{-1} to medium 7 × L 16 (385 ng Se · l^{-1}, 2 meq HCO_3 · l^{-1}) increased the level of F from 1 to 20–30 units compared to 0.2 to 10 units in the same medium without Se. Optimum B12 requirements were about 50–100 ng · l^{-1}, a concentration level suitable for applying a bioassay of this vitamin in fresh waters using the diatom as a test alga. Only a slight decrease of the algal biomass occurred above optimum up to 300 μg B12 · l^{-1} (Lindström, unpublished). Additions of B12 to medium 2 × L 16 (110 ng · l^{-1}) more than tripled the biomass yield and increased K_F ~ 7 times, but the vitamin effect on Se-deficient algae was only moderate (Fig. 5).

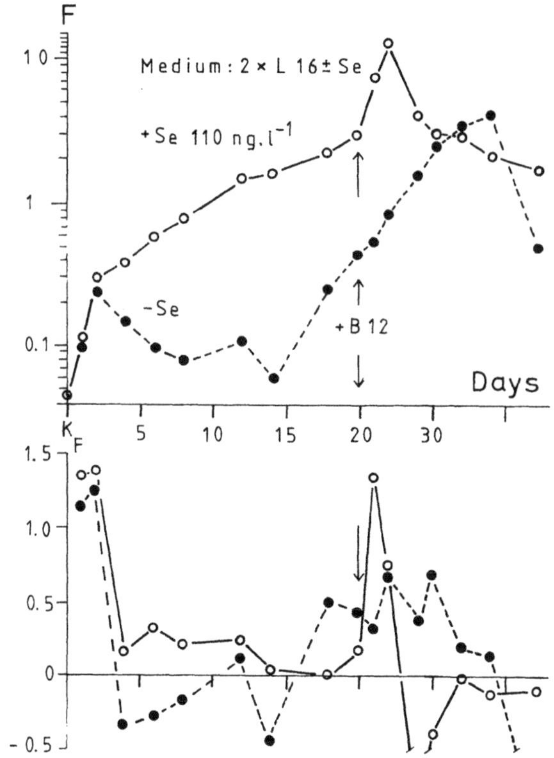

Fig. 5. Biomass development (F) and doublings · day^{-1} (K_F) of *Stephanodiscus hantzschii* var. *pusillus* grown in medium 2 × L 16 (110 ng selenite-Se · l^{-1}), and in the same medium without Se. Cultures from the experiment presented in Fig. 4 were spiked with vitamin B12 (100 μg · l^{-1}) after 20 days of growth.

An exploratory study of eight common representants of green algae showed significantly increased biomasses (25–67%) and growth rates (2–10%), when selenite-Se was added to the culture media (Table 3). The Se-stimulatory effect was evident after 6–8 days from the start of the cultures and lasted beyond their stationary phase (15–20 days). *Haematococcus pluvialis* Flotow, did not signifi-

Table 3. Biomasses (F) and doublings · day^{-1} (K_F) of eight representatives of green algae after various days of growth in batch cultures containing Lindström medium with and without addition of selenite-Se (two parallel experiments). The alkalinity was adjusted to 2 meq · l^{-1}.

Phytoplankton	Medium	Se ng · l^{-1}	F	SD	Day	K_F	SD	Growth period days
Chlorophyceae								
Chlamydomonas sp.	7×L 16	385	104	8	12	0.82	0.01	12
	7×L 16	0	74	6	12	0.78	0.01	12
Chlorella sp.	7×L 16	385[a]	109	71	12	0.94	0.08	12
	7×L 16	0	71	9	12	0.90	0.01	12
Dichtyosphaerium pulchellum Wood[b]	7×L 16	385	42	13	19	0.43	0.03	19
	7×L 16	385	27	0	19	0.39	0.00	19
Monoraphidium sp.	7×L 16	385	159	18	15	0.60	0.01	15
	7×L 16	0	115	1	15	0.57	0.00	15
Pandorina morum (Mull.) Bory	7×L 16	385	65	6	18	0.67	0.01	18
	7×L 16	0	51	1	18	0.65	0.00	18
Scenedesmus sp.	7×L 16	385	140	20	12	0.91	0.02	12
	7×L 16	0	111	1	12	0.88	0.00	12
Staurastrum sp.	7×L 16	385	15	3	18	0.48	0.02	18
	7×L 16	0	12	1	18	0.45	0.01	18
Tetraëdon caudatum (Corda) Hansg.	7×L 16	385	72	52	19	0.44	0.06	19
	7×L 16	0	43	21	19	0.40	0.03	19

[a] Enriched with extra 0.2 μg selenite-Se · l^{-1}, day 10.
[b] Axenic.

cantly respond to enrichments of Se and a dinoflagellate, *Gymnodinium inversum* Nyg. showed only ~5% ($N = 7$) biomass increase with additions of this element to the culture medium (not presented in Table 3). *Chlorella* sp. showed ~53% (day 12) to ~5% (day 18) more cells in the Se-enriched flasks compared to cultures without Se-addition. However, a protective effect of Se (Vokal-Borek 1979; Rudd *et al.* 1980), acting against inhibiting superoptimal nutrient concentrations cannot be excluded in the experiments with the green algae. The occurrence of impurities, bacteria and algal excreta may also effect the fluorescence yield.

If these data are generally valid, selenium may be characterized as an element with ecological relevance for a number of common phytoplankton genera which often dominate in fresh waters. They also emphasize the significance of inorganic and organic micronutrients, which beside PO_4-P and NO_3-N may limit algal growth in natural waters.

Se and algae in Lake Erken

In 1979, the algal succession in relation to various species of Se was investigated in the calciferous and mesotrophic Lake Erken.

Decreases in selenite-Se and bioactive Se coincided with increases in Tot-Se, bioinactive-Se and algae (Fig. 6). The temporal decrease in selenite-Se at a depth of 1 and 5 meters corresponded to a similar change in bioactive-Se at these depths when the dinoflagellate *Woloszynskia ordinata* reached maximum biomasses (6 mg · l^{-1}) at the surface in the middle of March. These results indicate a rapid algal assimilation of the dominating selenite-Se fraction of the bioactive-Se. However, at the end of the declining phase which occurred in the beginning of April during spring flow and high precipitation, the bioactive Se fraction was rich probably in organic Se from runoff water and cell excreta, and reached its highest values, ~60–80 ng · l^{-1}. At the same time there were only small amounts of selenite-Se, <30 ng · l^{-1}, remaining in the water column.

The dinoflagellate *Peridinium aciculiferum* and the diatom *S. hantzschii* var. *pusillus* dominated the algal bloom at the end of April. During this period the bioactive Se, probably dominated by organic Se, selenate-Se and to a lesser extent by selenite-Se, declined to values near zero with almost no detectable selenite-Se, while the bioinactive Se content increased (~40 ng · l^{-1}). These data point to a rapid algal uptake and release of the various

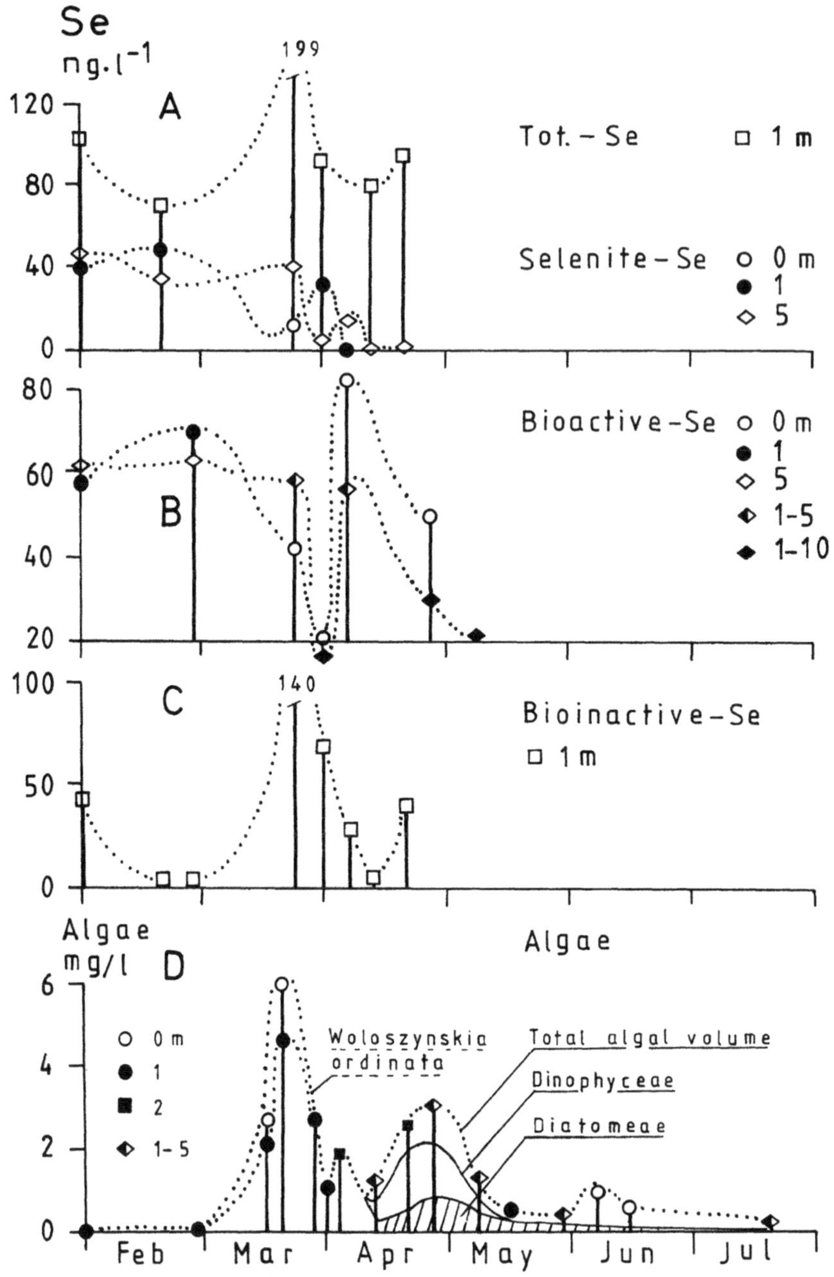

Fig. 6. Temporal variation of Tot-Se and selenite-Se (A), bioactive-Se (B), bioinactive-Se (C) and algae (D) in Lake Erken 1979.

forms of Se thus indicating its metabolic importance, where it may be considered as a nutrient factor important for the growth of certain phytoplankton species.

The results from culture experiments with algae, presented in Figs. 1–5 and Tables 2–3, further stress the ecological significance of Se and demonstrate it as an essential or growth-enhancing micronutrient for 10 species of algae in batch cultures.

Bioactive Se and pH

Table 4 illustrates the effect of pH on concentrations of bioactive-Se in surface water of acidified,

Table 4. Bioactive selenium (ng · l^{-1}) in water from acidified and lime-enriched forest lakes in the southwest and the northwest of Sweden. Sampling was performed during June and July 1979. The concentration of selenium was determined after 25 and 37 days of *P. cinctum* growth.

Lake	Depth	pH	Lime-enriched	Bioactive Se (ng · l^{-1})			SD
				25 days	37 days	$\bar{x}$	
Horsika	0	4.5	–	11	18	15	5
Rävekärr	0	6.5	yes	65	78	72	9
Gårdsjön	0	4.7	–	18	19	19	1
Västra Röd-vattnet	0	7.1	yes	40	33	37	5
Östra Röd-vattnet	0	5.0	–	14	10	12	3
Årsjön	1	5.7	–	22	25	24	2
Långsjön	1	6.6	yes	25	32	29	5
Upprämen	0.5	5.2	–	21	20	21	1
Upprämen	3	5.2	–	21	22	22	1
Skirsjön	0	6.1	fertilized with urea and NH_4NO_3	81	93	87	8
Lilla Dröpplan	0	5.6	humic	63	65	64	1

calcium-enriched, fertilized and humic waters in the western part of Sweden. Calculation of the regression coefficient (*r*) of bioactive Se (ng · l^{-1}) in relation to the pH status of these lakes, except Rävekärr, Skirsjön and Lilla Dröpplan, showed an apparent positive correlation ($N = 7$; $r \sim 0.92$; $pH = 0.107 \times$ ng Se · $l^{-1} + 3.1$), indicating much lower bioassimilable levels of Se in the acidified lakes (pH < 6) than in the lime-enriched or calcium-rich lakes (pH > 6). The lime-enriched Rävekärr as well as the fertilized Skirsjön and the humic Lilla Dröpplan demonstrated bioactive Se concentrations of more than 60 ng · l^{-1}.

These results indicate a deficiency of bioactive forms of Se in acidified waters, which may limit the growth of certain species of phytoplankton.

Discussion

Criteria for the establishment of a micronutrient according to Arnon & Stout (1939) are that: 1) a deficiency makes it impossible for the plant to complete its life cycle; 2) the deficiency is specific to the element; 3) the element participates directly in the nutrition of the plant.

These criteria seem valid for Se and the growth of *P. cinctum* and are for the present only tentatively suggested for some of the other algae in this investigation.

The bioactivity of Se concerning the response of *P. cinctum* (Figs. 1–2) indicates selenite-Se and organic Se to be more readily assimilable than selenate-Se.

Maximum content, measured chemically, of Se in *P. cinctum*, cultured in medium 2 × L 16 (Table 1) is up to 150×10^{-6} ng · $cell^{-1}$ (M. Remberger, pers. comm.) and was measured after about 20 days from the start of the cultures (the period of maximum growth rate of the alga; Lindström, in prep.). The content of Se per cell, based on cell numbers at optimum selenium concentration during maximum growth rate was $36 \cdot 10^{-6}$ ng per cell. In Se-depleted cells, cultured in the same medium without Se, the cell content was only 20×10^{-6} ng · $cell^{-1}$ (M. Remberger, pers. comm.). However, growth continues with only $5 \cdot 10^{-6}$ ng Se · $cell^{-1}$. The high cell content relative to the medium concentration (110 ng · l^{-1}) indicates an active uptake of the element by this alga.

Se-saturated cells may grow well for some time (0–15 days) even at concentration levels of 2–5 ng Se · l^{-1}. However, during the lag phase, until the cultures reach the exponential phase when maximum doublings day^{-1} (k_{max}) is reached, a stored pool of cellular surplus Se is probably utilized as with PO_4-P (Lindström 1982). An equilibrium is thus established between cell content and uptake rate which is determined by the medium concentration. The cells are thus not exactly in the same growth phase when comparing the responses at various nutrient concentrations.

Obtained values of Se optima and optima of, e.g., PO_4-P generally also will increase with time due to nutrient consumption causing the initial ambient concentrations to be successively higher than the real concentrations, especially during the stationary growth phase. The irregular growth of the alga during the lag phase (op. cit.), which also is prolonged in Se-depleted cells, external nutrient decrease during the growth procedure, trace contaminants and the effects of various metabolic states of the inocula cells may limit the batch technique to indicate suitable ranges of nutrient concentrations for optimum growth. Chemostate techniques may solve these problems although difficulties may arise with the negative effects of turbulence, cell wall production and the maintaining of axenic conditions.

The generally less efficient utilization of selenate-Se relative to selenite-Se by *P. cinctum* (Fig. 1) is presumably explained by the additional energy-consuming metabolic steps required for the alga to reduce and incorporate selenate into amino acids (Se^{-II}). The prolonged lag phase of the alga when fed on selenate-Se may also indicate an adaptative process or a shift to other metabolic pathways. The lag phase may be shortened if only selenate-Se is given to the alga during several inoculations.

The high bioavailability of selenite-Se generally agrees with the findings of Butler & Peterson (1967), who demonstrated that selenite-Se was absorbed and quantitatively assimilated by *Spirodela* approximately three times more efficient than selenate-Se. However, the incorporation of selenite-Se into *Scenedesmus dimorphus* was only 2% of that of Se-methionine (Sandholm *et al.* 1973). But Table 3, however, illustrates that the growth of *Scenedesmus quadricauda* was obviously improved by low additions of selenite-Se (385 ng · l^{-1}). Such differences between the algal species in the uptake capability are less significant when considering the very low concentrations (<1 μg Se · l^{-1}) that seem to be required for the stimulation of the algal growth (Figs. 1–5, Tables 2–3).

Shrift (1973) showed that organic Se (Se-methionine) was available for *Chlorella vulgaris* and that the uptake was active, but inhibited cell division. With addition of organic Se both *P. cinctum* and *S. hantzschii* var. *pusillus* reacted with improved growth, while regarding selenite-spiked *Chl. vulgaris* the growth response was slightly stimulated (Tables 2–3).

An unpleasant odour, presumably from sulphur compounds or from volatilized methylated Se compounds, was recognized in the *Stephanodiscus* cultures when grown on organic Se (Table 2). In this case a biomethylation of Se seems possible where B12 coenzymes are involved (Ridley *et al.* 1977). The biomethylation of Se is considered one of the metabolic pathways for environmental cycling of the element (Gamboa Lewis 1976).

Tetraselmis tetrathele and *Dunaliella minuta*, two marine algae, transferred selenite-Se similar to the metabolic pathways of sulphur. The extracted Se compounds were identified as Se analogs of sulphur containing acid derivates, and H_2Se played only a minor role in the algal metabolism (Wrench 1978).

Depending on the algal species and the form of Se, various uptake mechanisms of this element, although not always directly comparable to those of the sulphur, are presumably operating independent of whether the alga requires the element for growth or not.

Escherichia coli and a number of anaerobic bacteria are known to possess the selenium-dependent form of formate dehydrogenase (Stadtman 1980). According to Stadtman (op. cit.) the formate dehydrogenase of *E. coli* contains Se, Mo and Fe and is integrated in its nitrate reductase system. Other Se-dependent enzymes are glycine reductase (Turner & Stadtman 1973), nicotinic acid hydroxylase (Imhoff & Andreesen 1979), xanthine dehydrogenase (Wagner & Andreesen 1979), thiolase (Hartmanis 1980) and seleno-tRNAs (Chen & Stadtman 1980). The existence of algal enzyme equivalents may not be excluded. The high Mo and Fe requirements of *P. cinctum* (~20 μg · l^{-1}), compared to other micronutrients (Table 1), the high affinity to both inorganic and organic Se (Figs. 1–2) and the efficient utilizing of NO_3-N (Lindström, in prep.) may thus point to a possible participation of a Se-active enzyme in the nitrate reductase system of the alga.

Enlarged cells of *Chlorella* (Schrift 1973) and inhibition of apical structure development in *Acetabularia* (op. cit.) are other reported effects of Se when added to nutrient media. Compared to these findings Se starvation of *P. cinctum* inhibited the cellular proliferation and increased its cell diameter up to 17%, indicating the role of this element in cell division (Lindström, in prep.). In addition, its function as an indispensable nutrient factor for obtaining high biomasses, growth rates (Figs. 1, 2), final yields and oxygen production was also distinguished (Lindström & Rodhe 1978). Similar metabolic use of the element seems also plausible for *S. hantzschii* var. *pusillus* (Figs. 3–5) and possibly for certain green algae (Table 3). These results are in line with those of Pintner & Provasoli (1968) who found that 100 ng · l^{-1} to 10 μg · l^{-1} of Se as H_2SeO_3 stimulated growth of some marine chrysomonads belonging to the genera of *Chrysochromulina* and *Dicrateria,* although the dose-response curve was uneven.

In alkaline waters the selenite-Se fraction decreases very rapidly by algal uptake (Fig. 6). But under acidic conditions, like in some of the lakes presented in Table 4, decreased levels of dissolved Se has been noted for river waters, probably due to the formation of insoluble Se complexes with iron and iron hydroxides (Measures & Burton 1978). These observations also agree with the general incorporation characteristics of selenite-Se into agricultural crops, which attributes an increased uptake of the element to a high pH in the soil (Gissel-Nielssen 1971).

The bioassay of Se, performed under alkaline conditions (pH ~ 8) during >20 days, possibly overestimates the content of bioactive Se in acid test waters because of an oxidation of elemental Se or of metal selenides during the assay period. This process may be slow. However, biologically it might also take place more rapidly within the algal cells. Such an incorporation was observed in *Spirodela oligorrhiza* (Butler & Peterson 1967). An atmospheric oxidation of the colloid before entering the plant was not excluded in this case.

Several marine algae need vitamin B12 for their growth (Provasoli 1963), but regarding freshwater diatoms such a requirement is reported only from Lake Washington and two Japanese lakes, where the concentration dropped to low values (~4 ng · l^{-1}) during the season when *Cyclotella* species developed (Swift 1980). A specific B12 requirement was apparent in *S. hantzschii* var. *pusillus* (Fig. 5). This vitamin effect may act synergistically with Se.

The capability of *S. hantzschii* var. *pusillus* to assimilate and utilize organic Se (Table 2) and vitamin B12 for growth (Fig. 5) may explain its abundance in Lake Erken although almost no selenite-Se was detectable in the water. The ecological consequences are interesting because its appearance coincided with the decline of *Woloszynskia ordinata,* when much bioactive Se, including organic Se, and probably vitamins were released into the water (Fig. 6). The organic Se may constitute up to 50% of the dissolved Se (Hesslein *et al.* 1980; Uchida *et al.* 1980). However, almost no organic Se was found in Japanese rivers (Suzuki *et al.* 1981).

The known properties of selenite to adsorb and form colloids with iron compounds (bioinactive Se) at low pH may explain the decreased bioactivity of this element in acidified lakes (Table 4). Other mechanisms such as chemical and microbial reduction to various selenides and formation of elemental Se may further reduce its bioavailability in such waters. The deficiency of bioactive Se in acid waters is probably also amplified by coincident algal consumption. By adding lime or fertilizers to acidified lakes the pH, biological activity, and bioavailability of Se increased (Table 4), presumably due to formation of selenites and selenates of, e.g., Ca.

If Se is involved as a growth factor in the nutrient metabolism of algae, the current acidification of lakes in some instances may lead to a depletion of the bioavailable pool of this element, limiting certain algae and further enhancing the general algal nutrient deficiency in acidified waters.

The chelating capacity of humic colloids, containing Se, and their direct incorporation into the algal cells, may explain the higher bioactivity of Se in the humic lake in spite of its low pH status (Table 4).

Certain forms of dinoflagellates like *P. cinctum, P. willei, P. pusillum, Ceratium carolinianum* (Bailey) and of diatoms like *Fragilaria* sp. and *Tabellaria* sp. have been observed in acidified waters (Morling, pers. comm. 1980). They are presumably favoured by a reduced competition for light and nutrients and by a lowered grazing pressure by herbivores, and might rapidly incorporate the available Se. If Se-requiring dinoflagellates do occur in acid waters, they may yield considerable biomasses even at low Se concentrations when other nutrients are in surplus. *P. cinctum* fa. *westii* was found to produce 4 000 cells · ml^{-1} at a concentration of only 10–20 ng Se · l^{-1} (Lindström, in prep.). Despite the chemical and the biological elimination of the bioactive Se, such a concentration level and more is probably temporarily maintained in the lakes of acidified areas through high and frequent precipitation, which may contain high amounts of Se from fossil fuels (Traversy *et al.* 1975). Such conditions might lead to an increased production of, e.g., Se-requiring algae if other nutrients are sufficiently available.

Precipitation and runoff are important sources of Se transport into lakes. Selenates, selenites and organic Se (bioactive Se) are easily incorporated into primary producers and thus transported to the sediments in seston, as colloids associated with iron or ironhydroxides, or as metal selenides and elemental Se (bioinactive Se). Selenium compounds may also be volatilized into the atmosphere via biomethylation.

The bioactive Se is stable under alkaline and oxygen-rich conditions, but low pH and a reducing environment favour the existence of the various bioinactive forms of Se.

The element is dietarily important for mammals and essential for certain bacteria, but also an ecological role for single-celled plants seems relevant. In Sweden where a geological deficiency of Se is established, the element may sometimes be a metabolic limiting or a critical nutrient factor for growth of certain plankton algae.

The metabolic roles of Se and many other trace elements in fresh waters are little known and almost unexplored. A better understanding of their functions and bioavailability under various ecosystem conditions may also contribute in explaining some of the irregularities found in algal macronutrient utilization and cycling of, e.g., phosphorus and nitrogen.

Acknowledgements

I wish to thank C. Eriksson for her assistance with the bioassays, M. Remberger for kindly contributing the chemically measured data on selenium, P. Blomqvist for counting of phytoplankton, G. Ahlgren, C. Forsberg, and L. Fries for critically reading the manuscript.

References

Andreesen, J. R. & Ljungdahl, L. G., 1973. Formate dehydrogenase of Clostridium thermoaceticum: incorporation of selenium-75, and the effect of selenite, molybdate, and tungstate on the enzyme. J. Bact. 116: 867–873.

Andreesen, J. R. & Ljungdahl, L. G., 1974. Nicotinamide adenine dinucleotide phosphate-dependent formate dehydrogenase from Clostridium thermoaceticum: purification and properties. J. Bact. 120: 6–14.

Arnon, D. I. & Stout, P. R., 1939. The essentiality of certain elements in minute quantity for plants with special reference to copper. Plant Physiol. 14: 371.

Broyer, T. C., Lee, D. C. & Asher, C. J., 1966. Selenium nutrition of green plants. Effect of selenite supply on growth and selenium content of alfalfa and subterranean cover. Plant Physiol. 41: 1425–1428.

Butler, G. W. & Peterson, P. J., 1967. Uptake and metabolism of inorganic forms of selenium by Spirodela oligorrhiza. Aust. J. biol. Sci. 20: 77–86.

Chen, C.-S. & Stadtman, T. C., 1980. Selenium containing tRNAs from Clostridium sticklandii: cochromatography of one species with L-prolyl-tRNA. Proc. Natl. Acad. Sci. U.S.A. 77: 1403–1407.

Committee on Medical and Biologic Effects of Environmental Pollutants, 1976. Selenium. Natl. Acad. Sci., Wash. D.C., 203 pp.

Eriksson, C., 1981. Effects of selenium and phosphorus on growth of the diatom Stephanodiscus hantzschii v. pusillus. In: Håkansson, H. (ed.). Rapport från diatomésymposium i Lund, maj 1981. Rep. 22: 83–102.

Gamboa Lewis, B.-A., 1976. Selenium in biological systems, and pathways for its volatilization in higher plants. In: Nriagu, J. O. (ed.). Environmental Biogeochemistry, 1, pp. 389–409. Ann Arbor Science, Michigan, U.S.A.

Gissel-Nielsen, G., 1971. Influence of pH and texture of the soil on plant uptake of added selenium. Agric. Food Chem. 19: 1165.

Gissel-Nielsen, G., 1977. Control of selenium in plants. Risø Rep. 370. Agric. Dep., Roskilde, Denmark, 42 pp.

Hartmanis, M., 1980. A new selenoprotein from Clostridium kluyveri that copurifies with thiolase. Fed. Proc. 39: 1772.

Hesslein, R. H., Broecker, W. S. & Schindler, D. W., 1980. Fates of metal radiotracers added to a whole lake: sediment-water interactions. Can. J. Fish. aquat. Sci. 37: 378–386.

Imhoff, D. & Andreesen, J. R., 1979. Nicotinic acid hydroxylase from Clostridium barkeri: selenium-dependent formation of active enzyme. FEMS Microbiol. Lett. 5: 155–158.

Lewis, B., Johnson, C. M. & Broyer, T. C., 1974. Volatile selenium in higher plants. Pl. Soil. 40: 107–118.

Lindström, K., 1980. Peridinium cinctum bioassays of Se in Lake Erken. Arch. Hydrobiol. 89: 110–117.

Lindström, K., 1982. Environmental requirements of the dinoflagellate Peridinium cinctum fa. westii. Abstr. Upps. Diss. Fac. Sci. 646, 23 pp.

Lindström, K. & Rodhe, W., 1978. Selenium as a micronutrient for the dinoflagellate Peridinium cinctum fa. westii. Mitt. int. Ver. Limnol. 21: 168–173.

Measures, C. I. & Burton, J. D., 1978. Behaviour and speciation of dissolved selenium in estuarine waters. Nature 273: 293–295.

Measures, C. I. & Burton, J. D., 1980. Gas chromatographic method for the determination of selenite and total selenium in sea water. Anal. chim. Acta 120: 177–186.

Measures, C. I. & Burton, J. D., 1980. The vertical distribution and oxidation states of dissolved selenium in the northeast Atlantic Ocean and their relationship to biological processes. Earth planet. Sci. Lett. 46: 385–396.

Moxon, A. L. & Olson, O. E., 1974. Selenium in agriculture. In: Zingaro, R. A. & Cooper, W. C. (eds.). Selenium, pp. 675–707. Van Nostrand Reinhold Company, New York.

Parker, C. A. & Harvey, L. G., 1962. Luminence of some piazselenols. A new fluorometric reagent for selenium. Analyst, Lond. 87: 558–565.

Pinsent, J., 1954. The need for selenite and molybdate in the formation of formic dehydrogenase by members of the Coli aerogenes group of bacteria. Biochem. J. 57: 10–16.

Pintner, I. J. & Provasoli, L., 1968. Heterotrophy in subdued light of 3 chrysochromulina species, pp. 25–31. Haskins Laboratories, New York.
Provasoli, L., 1963. Organic regulation of phytoplankton fertility. In: Hill, M. N. (ed.). The Sea 2, pp. 165–219. Blackwell Scientific Publications, London.
Ridley, W. P., Dizikes, L. J. & Wood, J. M., 1977. Biomethylation of toxic elements in the environment. Science 197: 329–332.
Robinson, W. O., 1933. Determination of selenium in wheat and soils. J. Assoc. offic. agric. Chem. 16: 423–442.
Rudd, J. W. M., Turner, M. A., Townsend, B. E., Swick, A. & Furutani, A., 1980. Dynamics of selenium in mercury-contaminated experimental freshwater ecosystems. Can. J. Fish. aquat. Sci. 37: 848–857.
Sandholm, M., Oksanen, H. E. & Pesonen, L., 1973. Uptake of selenium by aquatic organisms. Limnol. Oceanogr. 18: 496–498.
Schwarz, K. & Foltz, C. M., 1957. Selenium as an integral part of factor 3 against dietary necrotic liver degeneration. J. am. chem. Soc. 79: 3292–3293.
Selenium and Tellurium Abstracts. Selenium-Tellurium Dev. Ass., Conn., U.S.A.
Shrift, A., 1973. Metabolism of selenium by plants and microorganisms. In: Klayman, D. L. & Günther, W. H. H. (eds.). Organic Selenium Compounds: Their Chemistry and Biology, pp. 760–814. Wiley, New York.
Stadtman, T. C., 1974. Selenium biochemistry. Science 183: 915–922.
Stadtman, T. C., 1980. Biological functions of selenium. Trends in Biochemical Sciences (TIBS) 5: 203–206.
Suzuki, Y., Sugimura, Y. & Miyake, Y., 1981. Selenium content and its chemical form in river waters of Japan. Jap. J. Limnol. 42: 89–91.
Swaine, D. J., 1979. Biogeochemical cycling of mineral-forming elements. In: Trudinger, P. A. (ed.). Studies in Environmental Science 3, pp. 10–16. Elsevier Scientific Publishing Company, New York.
Swift, D. G., 1980. Vitamins and phytoplankton growth. In: Morris, J. (ed.). The Ecology of Phytoplankton, Studies in Ecology 7, pp. 329–368. Blackwell Scientific Publications, London.
Traversy, W. J., Goulden, P. D., Sheikh, Y. M. & Leacock, J. R., 1975. Levels of arsenic and selenium in the Great Lakes Region. Sci. Ser. 58. Inland waters directorate Ont. Reg., Wat. qual. Branch, Burlington, Ont., 18 pp.
Turner, D. C. & Stadtman, T. C., 1973. Purification of protein components of the clostridial glycine reductase system and characterization of protein A as a selenoprotein. Arch. Biochem. Biophys. 154: 355–381.
Uchida, H., Shimoishi, Y. & Toei, K., 1980. Gas chromatographic determination of selenium (-II,0), -(IV), and -(VI) in natural waters. Envir. Sci. Technol. 14: 541–544.
Vokal-Borek, H., 1979. Selenium. Inst. Physics. Univ. Stockholm, 112 pp.
Wagner, R. & Andreesen, J. R., 1977. Differentiation between Clostridium acidiurici and Clostridium cylindrosporum on the basis of specific metal requirements for formate dehydrogenase formation. Arch. Microbiol. 114: 219–224.
Wagner, R. & Andreesen, J. R., 1979. Selenium requirement for active xanthine dehydrogenase from Clostridium acidiurici and Clostridium cylindrosporum. Arch. Microbiol. 121: 255–260.
Watkinson, J., 1966. Fluorometric determination of Selenium in biological material with 2, 3, diamino naphthalene. Anal. Chem. 38: 92–97.
Wrench, J. J., 1978. Selenium metabolism in the marine phytoplankters Tetraselmis tetrathele and Dunaliella minuta. Mar. Biol. 49: 231–236.
Zingaro, R. A. & Cooper, W. C., 1974. Selenium. Van Nostrand Reinhold Company, New York, 835 pp.

Origin and production of phosphatases in the acid Lake Gårdsjön

Håkan Olsson
Institute of Limnology, Box 557, 751 22 Uppsala, Sweden

Keywords: forest water ecosystems, phosphatases, phosphorus, phosphatase budget, acid lakes

Abstract

The activity of acid phosphatases was followed for one year in Lake Gårdsjön as well as in the inlet and the outlet of the lake. A budget of the phosphatases was calculated, including an estimation of the production of phosphatases. The phosphatase activity was also measured in two basins upstream of L. Gårdsjön: the north basin and the south basin of L. Stora Hästevatten.

The acid phosphatase activity was very high compared with reported alkaline phosphatase activities in other lakes. About 95% of the phosphatases in L. Gårdsjön was produced in the lake, and the production was highest in early summer.

Small *Chrysophyceae* (<10 μm) probably produced the majority of the acid phosphatases in the investigated lakes, and accordingly could be favoured in environments with low phosphorus supply due to their ability to produce large amounts of phosphatases.

Introduction

Lake Gårdsjön, in southwestern Sweden, is an acidified and oligotrophic lake with very low phosphorus content. Phosphorus is also the nutrient which primarily restricts algal growth in the lake (Broberg & Persson 1981).

Very high acid phosphatase activity, compared with reported alkaline phosphatase activities in neutral lakes, existed in L. Gårdsjön (Jansson *et al.* 1981). Little information was found in the literature on the role of acid phosphatase (Jansson *et al.* 1981) and probably no further studies on phosphatases in acid lakes have been made. This justified a closer investigation of the origin, production and role of phosphatases in acid environments in the L. Gårdsjön area. In addition to this paper, two other parts of the study have been published (Jansson *et al.* 1981; Jansson 1981).

In Jansson *et al.* (1981) a characterization of the phosphatases by gel-filtration, the phosphatase activity at different pH and a preliminary budget of the phosphatase activity was presented. The phosphatases in L. Gårdsjön had a pH-maximum at 5.0 and the activity was inhibited by phosphate. At least four different types of acid phosphatases were found in the lake in July 1980. About 50% of the phosphatase activity was found in filtered water while 40% was found associated with seston smaller than 5 μm, including small algae (e.g. *Cromulina* and *Chlamydomonas*) and bacteria. Experiments to explain the extremely high phosphatase activity were presented by Jansson (1981). It was found that high concentrations of aluminum, a commonly observed feature of acid lakes, blocked the phosphorus esters in the lake, which was supposed to decrease the availability of phosphorus and in turn induced phosphatase production. Iron could probably have the same effect as aluminum, but existed in much lower concentrations and was therefore considered to be of less importance.

The present paper deals with phosphatases in

Hydrobiologia 101, 49–58 (1983).

three lakes, including variations during a whole year-cycle. The origin of the phosphatases and the regulatory factors of phosphatase activity is further analysed. A 'phosphatase budget' is presented for the whole investigation period. Production of phosphatases during different periods of the year was calculated from the budget and the connection between production, lifetime and activity of phosphatases is discussed.

Characteristics of the studied lakes

The acidified Lake Gårdsjön is situated in the southwest of Sweden, about 45 km north of Gothenburg (Fig. 1).

The bedrock in the drainage area consists of gneiss and granite, and often lacks soil-cover. The soil consists mainly of till. Coniferous forest dominates the vegetation.

Upstream of L. Gårdsjön there are three smaller lakes (Fig. 1). L. Stora Hästevatten, which is included in this investigation, is divided into two basins by a road embankment and will be considered as two lakes, being referred to as the north basin and the south basin.

Table 1 gives morphometric characteristics and Table 2 chemical characteristics of L. Gårdsjön and the two basins of L. St. Hästevatten. The water renewal time of L. Gårdsjön is 2.3–2.7 years. The lakes are clear, poor in nutrients and have low pH. The aluminum content is high.

Table 1. Morphometric data of Lake Gårdsjön and of the two basins of Lake Stora Hästevatten.

	Area (km^2)	Volume (10^6 m^3)	Mean depth (m)	Max. depth (m)
L. Gårdsjön	0.31	1.5	4.9	18.5
L. St. Hästevatten				
The north basin	0.03	0.1	3.2	6.5
The south basin	0.05	0.2	4.0	9.3

Material and methods

Sampling

Samples were taken in L. Gårdsjön and in the inlet and the outlet on 22 occasions from March

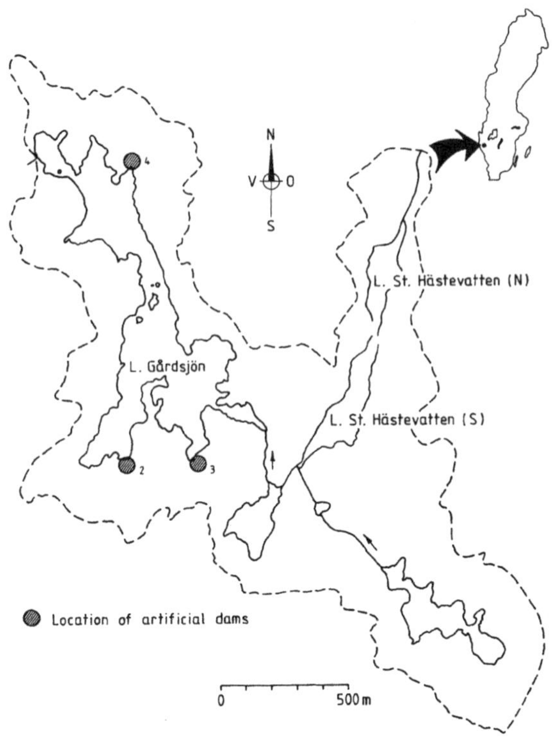

Fig. 1. The Lake Gårdsjön watershed.

1980 to May 1981. During summer-stratification samples were taken from three different strata (epi-, meta- and hypolimnion), and during winter from two strata (epi- and hypolimnion). When the lake was isothermal one sample was taken from the whole water-column.

Samples representing the water quality of diffuse inlets were taken on 18 occasions in three 'artificial dams' in the drainage area of L. Gårdsjön (Fig. 1). In both basins of L. St. Hästevatten, samples were taken from two strata during stratification and from the whole water-column during isothermal conditions.

Samples were frozen at −20 °C and were stored frozen until analysis after 8–219 days. On four occasions unfrozen samples were brought to the laboratory in Uppsala, immediately after sampling, for special studies.

Phosphatase activity assay

Phosphomonoesterase activity, abbreviated to 'phosphatase activity' in the text, was analysed at 20 °C with the substrate 4-methylumbelliferyl

Table 2. Water-chemical data of Lake Gårdsjön, (epilimnion) and the two basins of Lake Stora Hästevatten, (mean for the whole water-column). All concentrations are given in $\mu g \cdot l^{-1}$.

	L. Gårdsjön[a]	The north basin of[b] L. St. Hästevatten	The south basin of[b] L. St. Hästevatten
pH	4.6	5.3	5.2
Conductivity ($mS \cdot m^{-1}$)	6.7	5.7	5.5
Secchi disk transp. (m)	8.9	~6.5	6.9
Total-P	5.4	3.6	4.5
Part.-P	3.5	2.7	3.1
Total-N	397	300	367
$NO_2 + NO_3$-N	114	84	49
NH_4-N	48	18	33
Part.-N	56	–	–
Diss. org. C	2 100	2 540	3 881
Chlorophyll *a*[c]	1.05	0.70	1.25
Total-Al[c]	270	90	150

[a] Mean for the period 1979–1980.
[b] Mean for the period 5 February 1980–18 November 1980.
[c] Mean for the whole water-column 28 April 1980–15 October 1980.

phosphate (Boehringer Mannheim) as described by Jansson *et al.* (1981). Samples from L. St. Hästevatten were analysed at pH 5.2 and the other samples at pH 4.6 in a Tris-malate-acetate buffer (10^{-3} M with respect to each component). These pH-values were chosen in order to perform the analyses at a pH close to that in the analysed waters. The pH in the dams was about 4.15. The assay was conducted at pH 4.6 since the activity at the pH in L. Gårdsjön was the most relevant to budgeting phosphatases. The substrate concentration in the reaction tube was $1.1 \cdot 10^{-5}$ M.

The fluorescent product 4-methylumbelliferone was detected with a Turner 111 fluorometer equipped with Lamp No. 110–850, emitting UV light. As a primary filter Corning 7–60 was used, and Wratten 8 and 48 were used as secondary filters. A calibration curve (Jansson *et al.* 1981) was used to calculate the amount of product formed during the assay.

All activities in frozen samples were corrected according to the correction curve described below.

Effects of freezing

The effects of freezing on the phosphatase activity were investigated on 8 July 1980, 23 March (L. Gårdsjön), 25 March (L. Stora Hästevatten) and 18 May 1981. Phosphatase activity was analysed before freezing and at different time-intervals after freezing. The water was frozen in 20 ml aliquotes to avoid refreezing and thawing was done in a room-temperated waterbath.

In May 1981 tests were made on water from the metalimnion in L. Gårdsjön and the results were evaluated with non-linear least squares summary statistics. The curve obtained is presented in Fig. 2. When a sample was frozen at –20 °C for 20 days or more the phosphatase activity was reduced by about 60%.

No large differences were seen between different samples or in samples taken on different occasions.

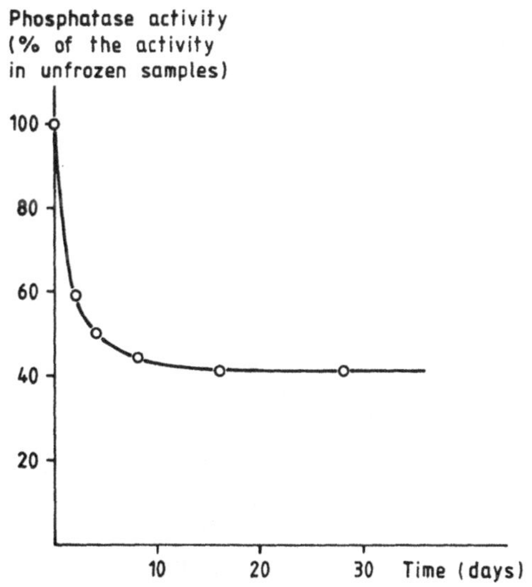

Fig. 2. The decline of phosphatase activity when samples were stored at –20 °C.

Therefore the curve presented in Fig. 2 was considered to be relevant as a correction curve for all frozen samples during the investigation.

Budget calculations

With the aim of obtaining information on the production of phosphatases in the lake, a factor describing the loss of phosphatase activity was introduced. The 'life-length' of phosphatases was studied in chloroform-treated water in July 1980 (Jansson *et al.* 1981). Data from that experiment were used to find an equation describing the losses of phosphatase activity with time by means of non-linear least squares summary statistics.

The 'decay curve' is presented in Fig. 3 and can be described with the equation

$$A_t = 100 \cdot (1 + a) \cdot (1 + a \cdot e^{b \cdot t})^{-1} \quad (1)$$

where A_t is the phosphatase activity at the time t in per cent of the activity at the time zero; t is the time in days after adding chloroform; and a and b are constants that were estimated to 0.076 (SD = 0.042) and 0.071 (SD = 0.012) respectively. This equation was used in a model to estimate the phosphatase production.

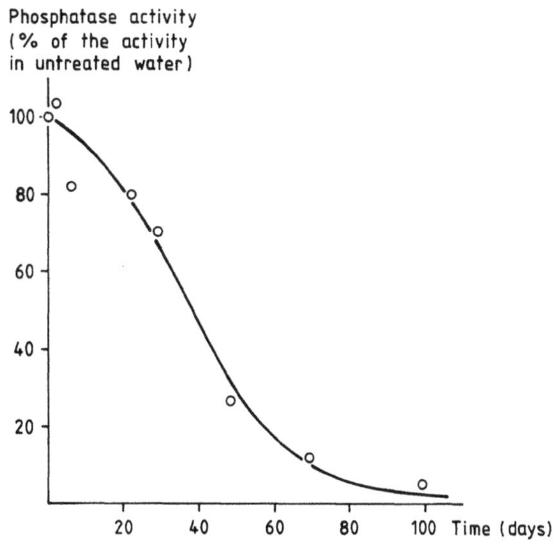

Fig. 3. The decline of phosphatase activity when adding chloroform to water from Lake Gårdsjön, 7 August 1980.

Transport of phosphatase activity has been calculated according to

$$A_i \cdot Q_i \quad (2)$$

for a certain time-period, i, of 14 days. A_i is the phosphatase activity in the middle of the period and Q_i is the accumulated water-flow during the same period. The phosphatase activity for periods with no sampling was extrapolated linearly from the two closest sampling events. The flows at the inlet and in the three dams have been calculated from the measured discharge in the outlet by factors proportional to the respective drainage areas. Phosphatase activity in precipitation was assumed to be zero.

The theoretical activity in the lake, after the period i, was calculated using the following equation:

$$\text{Theor}A_i = (V \cdot \text{Theor}A_{i-1} + A_{i,in} \cdot Q_{i,in} - A_{i,out} \cdot Q_{i,out}) \cdot V^{-1} \quad (3)$$

where A is the phosphatase activity and V is the lake volume. $A_{i,in} \cdot Q_{i,in}$ and $A_{i,out} \cdot Q_{i,out}$ represent the transport of phosphatase activity in the inflowing water and in the outlet according to (2), i indicating the number of the period.

For a certain period, the activity remaining from phosphatases produced and imported during previous periods, was calculated according to (1) and this activity was introduced as factor 'Theor A_{i-1}' in (3). The difference between the measured activity (mean for the whole water-body) and the theoretical activity gave the production of the phosphatases during the period. The model was run for 25 periods, from March 1980 to February 1981.

Results

Phosphatase activity budget and production of phosphatases

From March to June 1980 no large differences were seen between the accumulated transport of phosphatases into and out of L. Gårdsjön. Later during the year, especially during November-December, output was higher than input.

From March 1980 until January 1981, 9 150 mmol · min^{-1} of phosphatase activity had left and 5 900 mmol · min^{-1} had entered the lake, and during the same period, the accumulated inflow of phosphatases was 5% of the accumulated production of phosphatases. Without correction for losses of phosphatase activity, the contribution from the inlets was about 11% of the production.

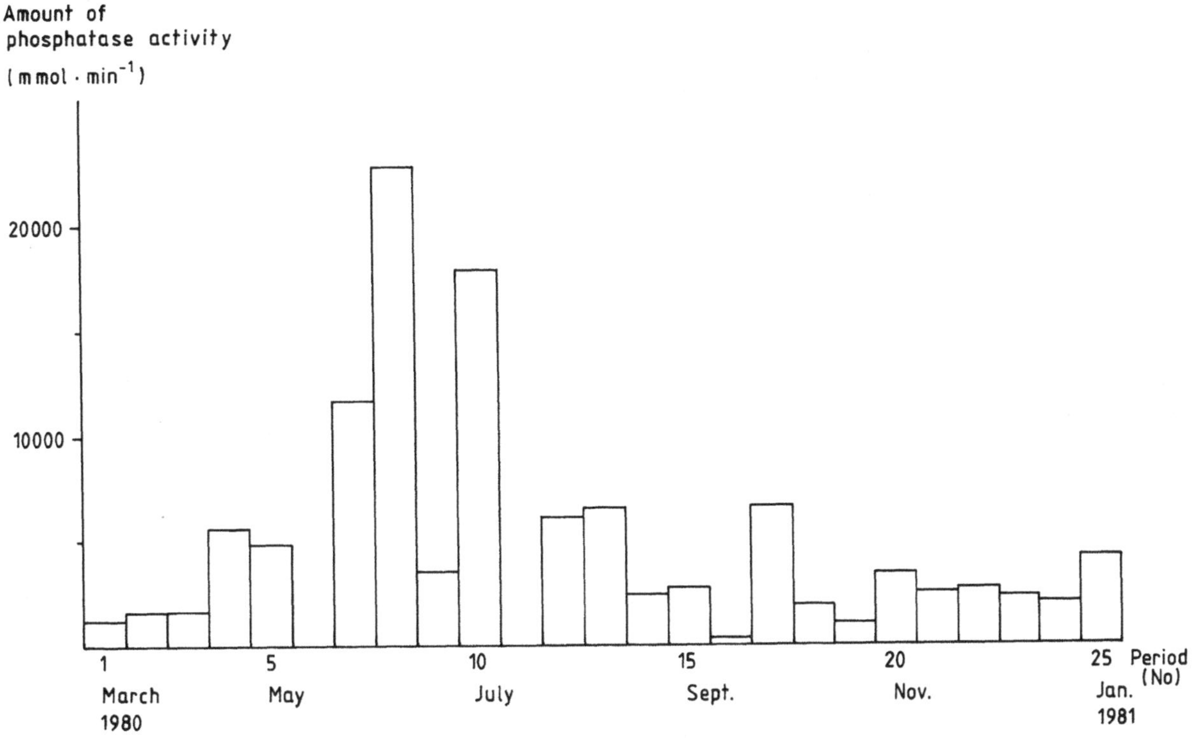

Fig. 4. The amount of phosphatase activity produced in Lake Gårdsjön during periods of 14 days, March 1980–January 1981.

The production of the phosphatases during each period of 14 days is shown in Fig. 4. The highest production (22 800 mmol · min^{-1}) occurred in early summer but there were large fluctuations during that time of the year. In periods 6 and 11 negative production values were achieved. These could be reduced or turned positive if the model was run with a higher 'decay rate' of the phosphatases.

The variations of the production of phosphatases broadly followed the variations of the phosphatase activity in L. Gårdsjön (see Fig. 5).

Seasonal and spatial variations of the phosphatase activity

The phosphatase activity in the studied waters is presented in Figs. 5–7.

In L. Gårdsjön and its outlet a peak in phosphatase activity was found from early June until late July. In the south basin of L. St. Hästevatten as well as in the inlet of L. Gårdsjön, a peak appeared half a month earlier, with the beginning in mid-May and the decline in early July.

In the dams and in the north basin of L. St. Hästevatten the phosphatase activity showed small variations over the investigated period. The lowest activities were found in the dams. The mean of the activity during the investigated period was 1.2 nmol · l^{-1} · min^{-1} (SD = 1.0) in the dams and 5.2 nmol · l^{-1} · min^{-1} (SD = 2.0) in the north basin of L. St. Hästevatten.

In the south basin of L. St. Hästevatten the activity was higher than in the north basin but the differences were small in autumn and winter. No summer peak was found in the north basin of L. St. Hästevatten. The north basin was isothermal during the whole summer, which contrasted to the south basin, where there was a well-developed thermocline. During the stratification period the activity was highest in the hypolimnion but vertical differences were generally small. In L. Gårdsjön the highest activity occurred in the metalimnion while the differences between the lower activities in the epi- and hypolimnion were small. The maximum activities in early summer exceeded the 'mean activities' during the rest of the year approximately 4–5 times in

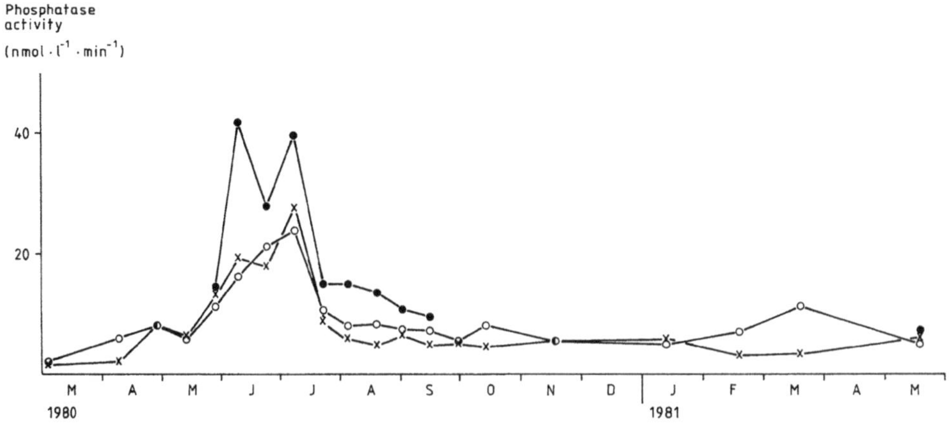

Fig. 5. Phosphatase activity in the epilimnion ○, the metalimnion ●, the hypolimnion x and the whole water-column ◐ of Lake Gärdsjön, March 1980–May 1981.

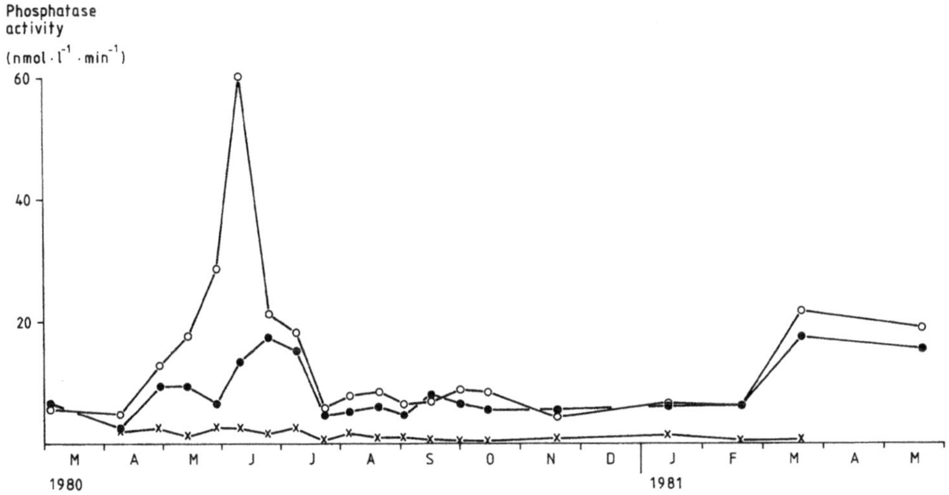

Fig. 6. Phosphatase activity in the inlet ○ and the outlet ● of Lake Gårdsjön and in the dams representing diffuse inlets x, March 1980–May 1981.

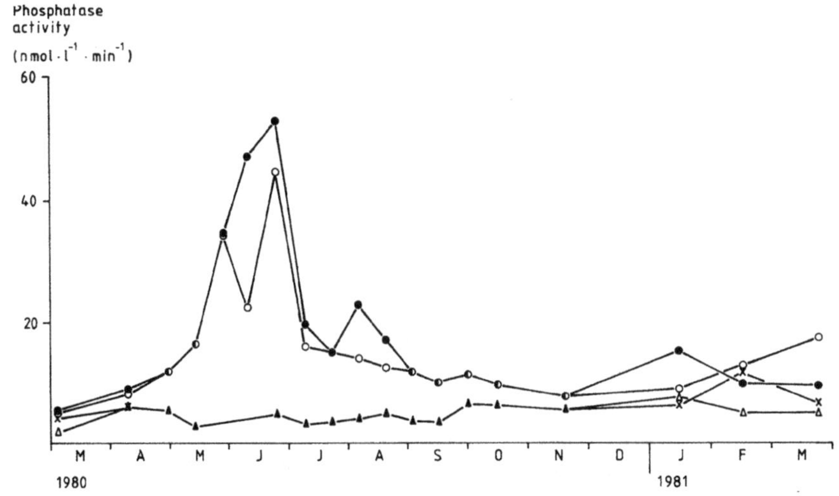

Fig. 7. Phosphatase activity in lake Stora Hästevatten, March 1980–March 1981. The epilimnion ○, the hypolimnion ● and the whole water-column ◐ of the south basin. The epilimnion x, the hypolimnion △ and the whole water-column ▲ of the north basin.

both L. St. Hästevatten and L. Gårdsjön.

In March and May 1981 high activities were measured in both the inlet and outlet of L. Gårdsjön. The high activity in the outlet is difficult to explain when comparing it with the activity in L. Gårdsjön. On no other occasion did a significantly higher activity appear in the outlet compared with the lake.

Phosphatase activity in relation to planktonic activity

In order to further elucidate the influence of different planktonic organisms on the production of phosphatases the seasonal variation of phosphatase activity was compared with different plankton parameters.

In the north basin of L. St. Hästevatten the phosphatase activity as well as chlorophyll concentration (Broberg, in prep.) was lower during summer compared with the other lakes. The peak in phosphatase activity in the south basin of L. St. Hästevatten coincided well with a peak in chlorophyll *a* but in L. Gårdsjön a chlorophyll *a* maximum appeared earlier in spring. In both lakes chlorophyll *a* was high in autumn but the phosphatase activity showed no increase at that time of the year.

Data on primary production in L. Gårdsjön and in the north basin of L. St. Hästevatten (Grahn, in prep.) did not correlate with the phosphatase activity. Measurements of primary production were not made in the south basin of L. St. Hästevatten.

Bacterial biomasses in L. Gårdsjön (Andersson, in prep.) were highest in May and in autumn 1980 and no covariation with phosphatase activity was seen. Bacterial countings were not made in L. St. Hästevatten.

As regards algal countings (Larsson, in prep.), total biomasses gave the same temporal variation as chlorophyll *a*. Small *Chrysophyceae* (<10 μm) dominated the algal biomass in the south basin of L. St. Hästevatten during the biomass peak in June. The biomass of small *Chrysophyceae* mostly covaried well with the phosphatase activity in the south basin of L. St. Hästevatten and in L. Gårdsjön (Fig. 8a and b). In June–July when the biomass of *Chrysophyceae* (<10 μm) declined rapidly, the decrease in phosphatase activity was not noticed until 14 days later.

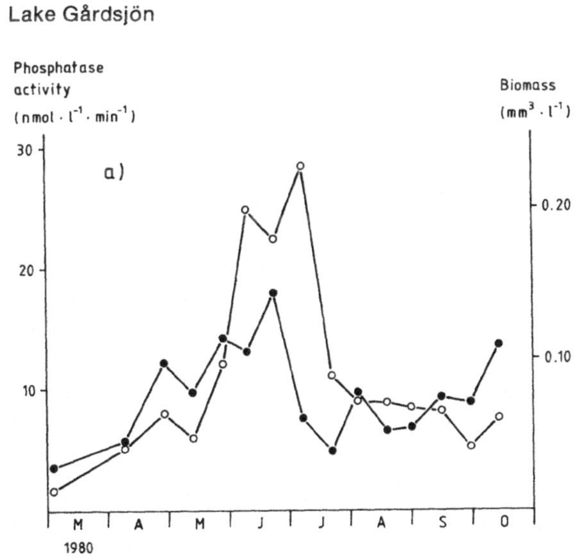

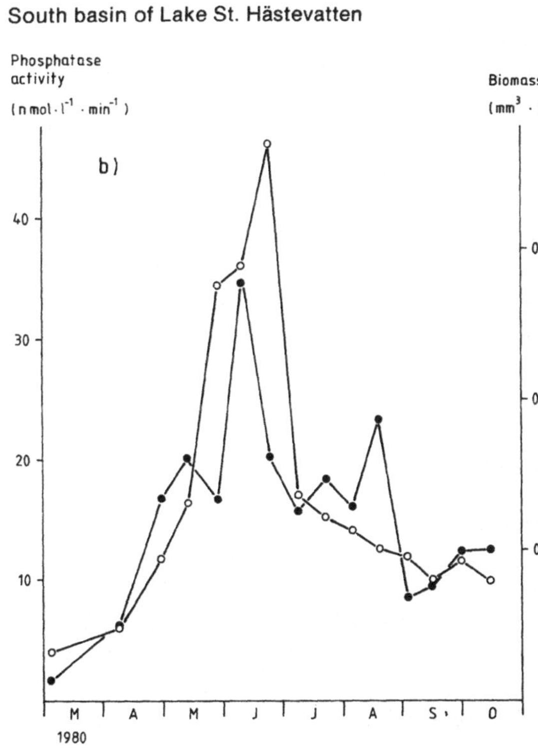

Fig. 8. Phosphatase activity ○ and biomass of *Chrysophyceae* (<10 μm) ●. Mean for the whole water-column.

Also vertical variations in phosphatase activity showed good agreement with vertical distribution of small *Chrysophyceae*. The highest phosphatase activity in the south basin of L. St. Hästevatten was found in the hypolimnion together with high biomass of *Chrysophyceae* ($<10\ \mu m$). In L. Gårdsjön the phosphatase activity, as well as the biomass of *Chrysophyceae* ($<10\ \mu m$), was highest in the metalimnion. In the hypolimnion the phosphatase activity was lower than expected from the biomass of *Chrysophyceae* ($<10\ \mu m$). This was most pronounced in early May.

In the north basin of L. St. Hästevatten the biomass of *Chrysophyceae* ($<10\ \mu m$) was lower during summer compared with the south basin. This was also the case with the phosphatase activity. No data from algal countings later than October 1980 were available when this paper was prepared.

Phosphatase activity in relation to phosphorus fractions and aluminum

No large differences in total and particulate phosphorus could be seen between the investigated lakes and there were no correlations, negative or positive, between phosphorus fractions and phosphatase activity.

Regarding the results from Jansson (1981) aluminum could be a factor regulating phosphatase activity in L. Gårdsjön. The aluminum concentrations during the period 28 April 1980–15 October 1980 in L. Gårdsjön, the south basin of L. St. Hästevatten and the north basin of L. St. Hästevatten were 270, 150 and 90 $\mu g \cdot l^{-1}$ respectively (Nilsson, in prep.), while the specific phosphatase activities (per biomass of small *Chrysophyceae*) were 160, 139 and 80 $nmol \cdot min^{-1} \cdot mm^{-3}$ respectively.

Discussion

Budgeting of phosphatases gives important information on the origin of the phosphatases in a lake. Stevens & Parr (1977) found that the inlets could contribute 40% to the increased alkaline phosphatase activity in Lough Neagh during the investigated period (February–June 1974). One possible reason for this high input of phosphatases was suggested to be sewage effluents.

There are no sewage effluents affecting L. Gårdsjön. The lakes in the area are acid and oligotrophic. High input of phosphatases could be due to high production of phosphatases in the lakes upstream of L. Gårdsjön and/or in the soil profile of the catchment area. However, the budget calculations in L. Gårdsjön showed that the majority of the phosphatases (~95%) were produced in the lake. As will be discussed below, there are results indicating that the losses of phosphatase activity have been underestimated, which means that the contribution from the inlets could be even less important compared with the production in the lake.

During conditions of low phosphorus supply many investigations have reported that the phosphatase activity increases due to derepression or induction of enzyme synthesis. Consequently, the phosphatase production may give valuable information about the phosphorus conditions. The measured phosphatase activity is dependent on the phosphatase production but there may be a considerable timelag between the two variables depending on the longevity of the phosphatases in the water. A great step forward in ecological studies would therefore be made if the phosphatase production could be determined.

An attempt to estimate the production was made when budgeting phosphatases in L. Gårdsjön. The results seem reasonable and show that the methods used are applicable. In two periods during the summer negative values of production were however obtained, probably due to the actual elimination rate of phosphatase activity during these periods being higher than that observed under laboratory conditions.

The laboratory experiments estimating the lifetime of the phosphatases did not simulate or control all factors which could affect the phosphatases in the lake. Losses due to sedimentation of seston containing phosphatases, increased amount of hydrolytical enzymes or chemical substances with the property to break down phosphatases, and increased sunlight causing breakdown of organic substances, are some factors not included in the laboratory experiment. Furthermore, the equation describing losses of phosphatase activity in the model theoretically could only be applied on newly produced phosphatases, while the laboratory experiment was made on lake water which contained phosphatases of different ages. If it had been possible to perform the laboratory experiment with newly produced phosphatases their lifetime would probably have been longer. However, the present results imply that of the factors not included in the experiment, those causing losses of phosphatases seemed to be most important. Since the estimated phosphatase production in different periods of the year gave very approximate values, seasonal changes of different parameters were only compared with the measured phosphatase activity.

The lakes with the lowest phosphatase activity in summer had the lowest biomass of phytoplankton, indicating that phytoplankton or bacteria associated with phytoplankton were the main producers of acid phosphatases. The pronounced phosphatase activity peak in L. Gårdsjön and in the south basin of L. St. Hästevatten coincided with the development of the biomass of *Chrysophyceae* (<10 μm) (Fig. 8a and b). Comparing the two variables in different strata of the lakes the results showed positive correlation. If it is assumed that small *Chrysophyceae,* at least during the summer, produced the same amount of phosphatase activity per biomass, the investigation indicates that a large part of the phosphatase activity in L. Gårdsjön and in L. St. Hästevatten came from algae among the group small *Chrysophyceae* (<10 μm). This is supported by results from the characterization of phosphatases in L. Gårdsjön with gel-filtration (Jansson *et al.* 1981). In July 1980 it was shown that the main part of seston-bound phosphatases were tied to particles passing a 5 μm filter. In this size fraction *Chrysophyceae* belonging to the genus *Chromulina* sp. were highly dominant.

Chrysophyceae have been shown to frequently dominate phytoplanktonic biomass in oligotrophic lakes (Schindler & Holmgren 1971; Ramberg 1979; Rosén 1981). *Chrysophyceae* was also shown to dominate in the eutrophic L. Erken during periods of low phosphorus supply, such as following the spring bloom (Pechlaner 1970; Pettersson 1980). Simultaneously the alkaline phosphatase activity was very high. Due to their ability to create a high phosphatase activity, *Chrysophyceae* could be favoured in environments of extremely low phosphorus availability.

In July the decline in phosphatases activity was delayed compared with the biomass of *Chrysophyceae* (<10 μm). A similar pattern was found by Petterson (1980) between the spring peak of chlorophyll *a* and the alkaline phosphatase activity in L. Erken in the years 1975–1978. Pettersson suggested the delay to depend on the algae containing surplus phosphorus in the beginning of growth and that the production of phosphatases increased when the surplus phosphorus was depleted after the biomass maximum.

In L. Erken the ratio surplus phosphorus/particulate phosphorus was at a minimum of about 0.07 in May and about 0.15 in April. In L. Gårdsjön the ratio was about 0.05 (mean) in 1979 with small variations (Broberg & Persson 1981). The ratio in 1980 should not be very different because phosphorus conditions were about the same as 1979. These figures indicate that the surplus phosphorus in L. Gårdsjön was constantly very low so the internal phosphorus supply should be depleted already at the onset of biomass increase. The maintained acid phosphatase activity in L. Gårdsjön can therefore hardly be explained as an effect of changed surplus phosphorus.

Considering the long 'survival' of phosphatase activity (Fig. 3) it is likely that the phosphatases can be active for 14 days after production. Since samples were taken fortnightly it is possible that the time-lag was shorter than 14 days. This means that the previously produced phosphatases could very well account for the observed delay. A long lifetime of alkaline phosphatases in lakes has been suggested by Healey (1973) but no information about lifetime of acid phosphatases in lakes has been found.

Jansson (1981) proposed that high concentrations of aluminum caused a high phosphatase activity in L. Gårdsjön. It was suggested that aluminum reacted with the phosphatase substrates and thereby protected the substrates against enzymatic hydrolysis. This inhibitory effect on recycling of phosphorus was compensated by an increased production of phosphatases.

A comparison between the investigated lakes, during the vegetation period, shows that the higher the concentration of aluminum the higher was the specific phosphatase activity (activity per biomass of small *Chrysophyceae*). If it is assumed that the lifetime of the phosphatases was the same in the three lakes it can be concluded that the highest phosphatase production took place in the lakes with the highest aluminum concentrations. My results thus support the experimental results of Jansson (1981) showing that aluminum inhibited recycling of phosphorus.

The main conclusions from this investigation are summarized below:

About 95% of the acid phosphatases in L. Gårdsjön were produced in the lake.

Small *Chrysophyceae* (<10 μm) were the major producers of acid phosphatases in the investigated lakes.

The determination of the production of phosphatases gave approximate values, but with a better estimation of the losses of phosphatases in the lake the method seems to be applicable in ecological studies.

Results from this investigation support the findings of Jansson (1981) that aluminum can be important in regulating the phosphorus availability in acid lakes. Further support for the effects of aluminum ought to be found from regional surveys of phosphatase activities and aluminum levels.

Acknowledgements

This investigation was carried out within the integrated 'Lake Gårdsjön-project' funded by the Swedish National Environmental Protection Board (SNV), The Fishery Board of Sweden and Swedish Council for Planning and Coordination of Research (FRN).

I thank the following members of the staff of the Lake Gårdsjön-project for supplying background data or technical help: S. Johansson and U. Nyström (hydrology), O. Broberg and G. Persson (phosphorus fractions, chlorophyll *a*, etc.), I. Nilsson (aluminum), P. Larsson (phytoplankton), A. Lydén and O. Grahn (primary production), I. Andersson (bacteria) and S. Larsson and P. Leyton (technical assistance).

I also thank B. Karlsson and G. Persson for their help in constructing a model for the budget calculations, and M. Jansson and K. Pettersson for valuable criticism of the manuscript.

References

Broberg, O. & Persson, G., 1981. Närsalttillgång i försurade sjöar. Fosfor, Kväve och organiskt kol i Gårdsjön. Mimeographed report, Inst. Limnol., Upps., Sweden. 55 pp. (in Swedish).

Healey, F. P., 1973. Characteristics of phosphorus deficiency in Anabaena. J. Phycol. 9: 383–394.

Jansson, M., 1981. Induction of high phosphatase activity by aluminum in acid lakes. Arch. Hydrobiol. 93: 32–44.

Jansson, M., Olsson, H. & Broberg, O., 1981. Characterization of acid phosphatases in the acidified Lake Gårdsjön, Sweden. Arch. Hydrobiol. 92: 377–395.

Pechlaner, R., 1970. The phytoplankton spring outburst and its conditions in Lake Erken (Sweden). Limnol. Oceanogr. 15: 113–130.

Pettersson, K., 1980. Alkaline phosphatase activity and algal surplus phosphorus as phosphorus-deficiency indicators in Lake Erken. Arch. Hydrobiol. 89: 54–87.

Ramberg, L., 1979. Relations between phytoplankton and light climate in two Swedish forest lakes. Int. Revue ges. Hydrobiol. 64: 749–782.

Rosén, G., 1981. Tusen sjöar. Växtplanktons miljökrav. Statens Naturvårdsverk Rapporter, Liber förlag, Stockholm. 120 pp. (In Swedish with English summary).

Schindler, D. W. & Holmgren, S. K., 1971. Primary production and phytoplankton in the Experimental Lakes Area northwestern Ontario, and other low-carbonate waters, and a liquid scintillation method for determining ^{14}C activity in photosynthesis. J. Fish. Res. Bd Can. 28: 189–201.

Stevens, R. J. & Parr, M. P., 1977. The significance of alkaline phosphatase activity in Lough Neagh. Freshwat. Biol. 7: 351–355.

Bacterioplankton in the acidified Lake Gårdsjön

Ingvar B. Andersson
Swedish Water and Air Pollution Research Institute (IVL), Box 5207, S-402 24 Gothenburg, Sweden

Keywords: forest water ecosystems, acidification, bacterioplankton, biomass, dynamic

Abstract

Total bacterial numbers in different strata of lake water and in inlet and outlet streams have been recorded during a yearly cycle. A calculated mean cell volume of 0.342 μm^2 has then been used to estimate bacterial biomass in the lake. Change of biomass during the year was substantial and the range was from about 0.1 g · m^{-3} to about 1.0–1.2 g · m^{-3}. The seasonal development included a spring–early summer increase followed by a decrease to the minimum in July–August. Correlation between epi- and hypolimnion was high and in both strata two dominant autumn peaks in biomass appeared. With the exception of the last autumn peak the development of bacterial biomass was closely related to development of phytoplankton biomass and production.

Introduction

The adverse effects of acidification to lakes and streams have received comprehensive attention in regions suffering from acid deposition. Many investigations and surveys have been performed to establish the effects and the extent of damage to aquatic biotas. Recent reviews are given by Overrein *et al.* (1980) and Muniz (1981). Investigations concerning bacterial numbers and biomass in oligotrophic lakes are still scarce and so far only a few studies have been made in acidified lakes (Thompson & Croll 1974; Traaen 1978, 1980).

The aim of the present study is to estimate bacterioplankton biomass and seasonal dynamics during acid conditions and the response after liming of the lake. Data presented here is from the first year.

Materials and methods

Lake Gårdsjön

Lake Gårdsjön is an oligotrophic acidified forest lake situated about 35 km north of Gothenburg on the Swedish west coast. The lake has been acidic since about 1965 and a seasonal acidification impact might have been obvious already in the late 1950s as in two other nearby lakes (Hultberg 1980). A general physico-chemical and biological description of Lake Gårdsjön has been given earlier (Grahn *et al.* 1974). Broberg & Persson (1981) give a description of vegetation, hydrology and morphometric data for the lake as well as a detailed balance of nutrient input and output.

The pH increased from winter to fall overturn, a general picture through all years of investigation (Fig. 1). This was most pronounced in near bottom waters, where even higher pH values (~5.5) have been recorded at the end of summer stagnation when anoxic conditions prevailed. The temperature and oxygen situation during 1980 (Fig. 2) also

Hydrobiologia 101, 59–64 (1983).

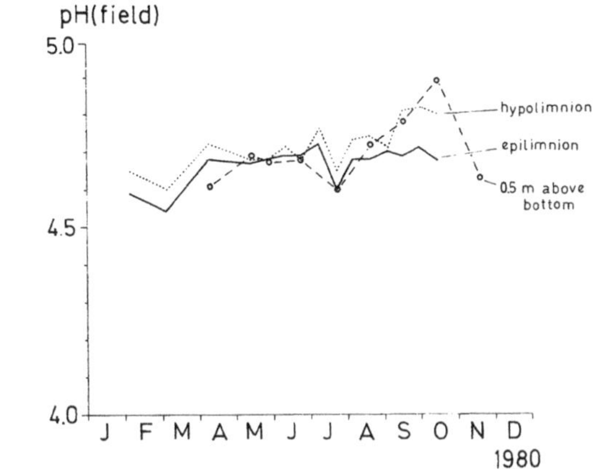

Fig. 1. The pH in Lake Gårdsjön during the study period.

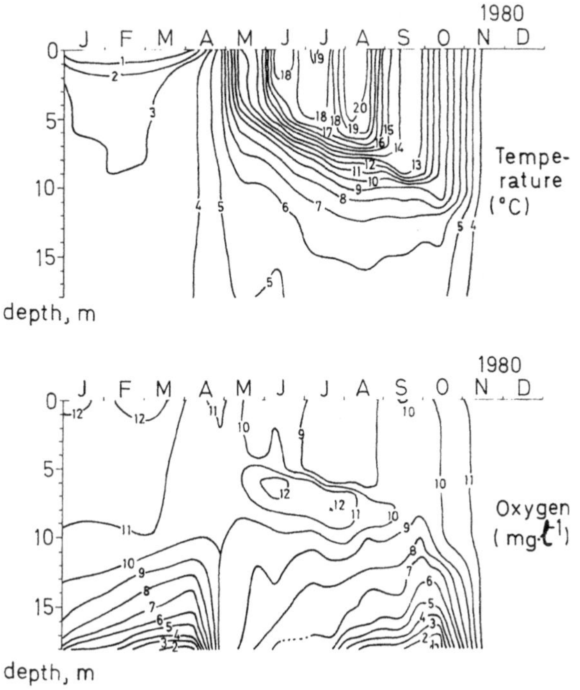

Fig. 2. Temperature and oxygen in Lake Gårdsjön.

agreed with earlier years. However, higher oxygen depletions in the hypolimnion have been recorded in 1975 and 1976.

Sampling

Large-volume nonsterile water samples were regularly taken at the greatest depth (integrated samples from epi-, meta- and hypolimnion), in inlet and outlet streams and in inlets from gauged micro-watersheds. In addition, supplementary samples in the lake were collected irregularly at specific depths. Subsamples of water for nutrient analysis were immediately fixed with $HgCl_2$-solution (Broberg & Persson 1981) and subsamples from these have been used for counting of total bacteria.

Microscopic preparation and examination

Counting was performed using epifluorescence microscopy after staining with acridine orange. The staining of filters and the solutions used followed Zimmerman *et al.* (1978). Duplicate 0.5-ml amounts were pressure-filtered onto polycarbonate filters (Nuclepore Corp.; 13 mm diameter; pore size 0.2 μm) and stained. In order to decrease background fluorescence, filters were rinsed by filtering twice with 2 ml of distilled water. All solutions were filtered (pore size 0.2 μm) at time of dispensing.

Stained filters were immediately mounted in immersion oil and examined with a Leitz Orthoplan equipped with a 200 W Hg-lamp. The filter set-up was a red-suppression filter (4 mm BG 38), exciting filters (2 $\times$ KP 490), and emission filters TK 510, K 515. Counting was usually performed through a 100X Planachromat oil immersion objective at a total magnification of 1 250. Numbers of bacteria were determined in ten or twenty rectangular fields of an ocular. Generally, this resulted in a total count of 200–400 bacteria. About 10% of the filters were also used for measuring and size classification (cf. Rheinheimer 1977). Both length and width of 100 randomly selected bacteria on each of these filters were measured by a calibrated ocular-micrometer. A weighted mean cell volume (0.342 μm^3) was then used to calculate the bacterial biomass, assuming a specific gravity of 1.0.

Results

Seasonal changes of bacterial numbers and biomass at the principal sampling point are shown in Fig. 3. In the epilimnion a large number ($3.6 \pm 0.3 \cdot 10^{12} \cdot m^{-3}$) and a high biomass ($1.2\ g \cdot m^{-3}$) were recorded in early February, while in the hypolimnion number and biomass were less than half. Biomass decreased substantially in the epi-

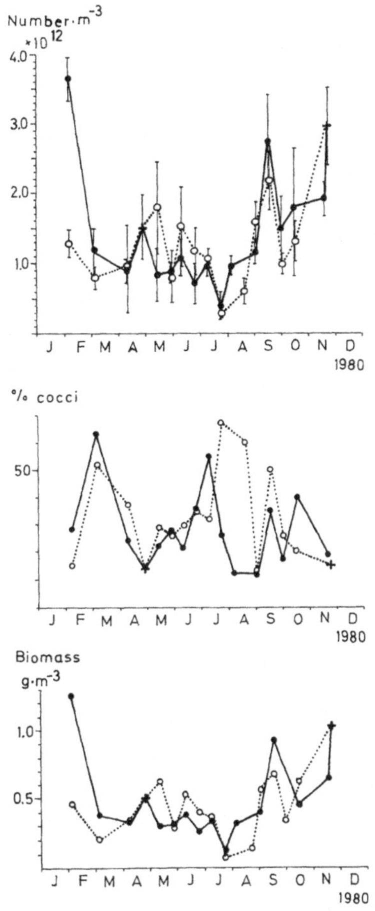

Fig. 3. Bacterial numbers, biomass and percent cocci in the epilimnion and hypolimnion (dotted line). Vertical bars denote 95% confidence interval.

limnion during February and was about equal to the hypolimnion biomass in late winter. At the time of spring circulation in late April, the total bacterioplankton biomass had again increased as shown by the integrated sample of the whole water column.

Peaks in the epilimnion and the hypolimnion differed only slightly. A somewhat larger biomass, 0.6 g · m^{-3}, was recorded in early May in the hypolimnion and after a second peak in June the biomass decreased to a minimum value of about 0.1 g · m^{-3} in July and August. After the early spring peak, biomasses in the epilimnion were generally smaller (about 0.3 g · m^{-3}) than in the hypolimnion. Also in the epilimnion a minimum (~0.1 g · m^{-3}) was recorded. Bacterial biomasses increased substantially during autumn and in the epilimnion

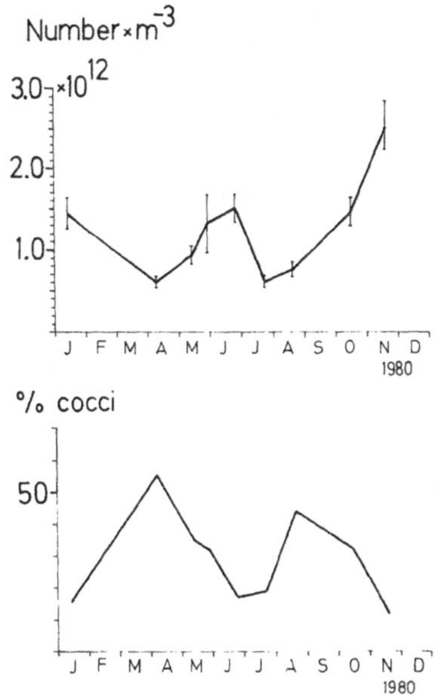

Fig. 4. Bacterial numbers and percent cocci in water samples from 0.5 m above bottom. Bars denote 95% confidence interval.

reached almost 1 g · m^{-3} in late September, with another peak of about the same size at fall overturn in November. The increase of biomass began earlier in the epilimnion than in the hypolimnion but both the peaks and decreases of biomass in September were simultaneous.

The numbers of bacterioplankton in the water layer close to the bottom (0.5 m above bottom) were in general similar to that in hypolimnion (Fig. 4). The increase in number during the summer occurred more slowly and was clearly delayed compared to the water masses above, with the maximum value in June. The minimum value in July coincided with that in epi- and hypolimnion. Also in the bottom water, the numbers increased notably during the autumn.

The percent cocci (Figs. 3, 4) showed a certain regularity, with low values during winter and in early summer and with peaks in spring and late summer. The highest percentages thus were recorded when total numbers of bacteria were lowest with the exception of the two hypolimnion samples from July and August. The same correlation was apparent for water from the inlet stream (Table 1). Numbers were at two occasions, early in winter and at time for snowmelt, about one order higher than usual and at both occasions the proportions of cocci were very low.

Table 1. Total number of bacteria (±95% confidence interval) and percentage of cocci in inlet and outlet streams and inlet from the gauged small drainage area No. 3.

Date	Number × $10^{12} \cdot m^{-3}$			% Cocci		
	Inlet	Outlet	Area No. 3	Inlet	Outlet	Area No. 3
1980						
01-07	16 ± 2.6	0.8 ± 0.4		6	51	
01-12	0.9 ± 0.3	1.3 ± 0.2	1.1 ± 0.2	32	18	35
02-04	2.0 ± 0.4	1.0 ± 0.2		51	29	
03-03	1.1 ± 0.3	1.6 ± 0.3		41	42	
04-08	12 ± 1.0	0.5 ± 0.2		12	55	
04-28	1.3 ± 0.2	0.8 ± 0.2	1.1 ± 0.2	27	57	55
05-12	0.9 ± 0.2	1.1 ± 0.2		25	7	
05-27	1.0 ± 0.3	0.6 ± 0.2		38	40	
06-09	2.0 ± 0.5			29		
06-23	0.9 ± 0.2	0.7 ± 0.2	0.8 ± 0.3	33	61	70
07-07	1.4 ± 0.3			31		
07-21	1.3 ± 0.3	1.1 ± 0.2		57	25	
08-04	3.0 ± 0.7	1.9 ± 0.3	2.3 ± 0.3	26	33	45
08-18	1.0 ± 0.3			22		
09-29	0.8 ± 0.5	1.4 ± 0.2		29	39	
10-13	1.7 ± 0.3	1.1 ± 0.3	1.1 ± 0.3	37	25	32
11-17	1.6 ± 0.2	1.2 ± 0.4	1.6 ± 0.3	33	43	39

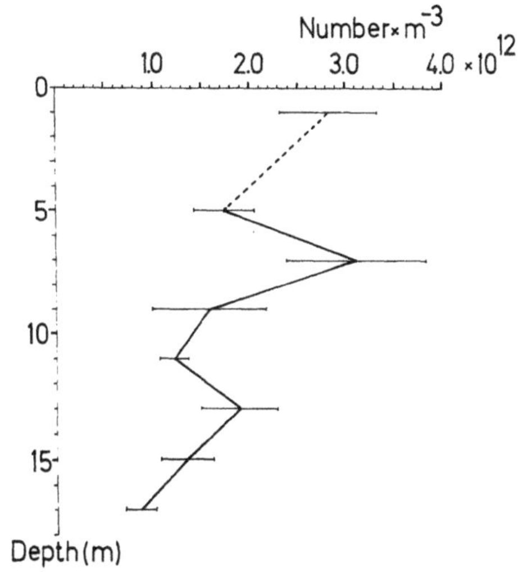

Fig. 5. Depth distribution of bacterial numbers in January 1980. The points are placed in the middle of the 2 m thick layers that were sampled. Bars denote 95% confidence interval.

A possible influence of bacteria from the inlet stream on numbers recorded at the sampling point, which was rather close to the inlet, can not be excluded based on available data. There are no figures on bacterial number in the inlet stream during maximum fall discharge 1979. However, a depth profile in the lake from early January 1980 (Fig. 5) showed higher numbers in the upper water layers. Since the layers examined were two meters thick, even higher numbers could have existed in the surficial layer. The high number observed in the epilimnion in February (Fig. 3) could well result from density stratified water from the inlet containing larger number of bacteria than in the lake water. The high number in the inlet water at snowmelt then might have been masked by mixing into the approximately homothermic water. During periods of high discharge in the fall of 1980 significantly increased numbers of bacteria were not recorded in the inlet stream.

Discussion

Lake Gårdsjön bacteria

Generally, rather large rod-shaped bacteria dominated the samples throughout the year. Cocci were a significant part of the total population only when total number was low. Since cocci mostly were very small the contribution to the biomass were almost insignificant (cf. Rheinheimer 1977; Stråskrabová & Komárková 1979; Salonen 1981). The percent cocci together with size distribution may, however, be indicative of successions or useful in comparing populations from different lakes (Zimmerman *et al.* 1978).

The weighted mean cell volume calculated for Lake Gårdsjön bacteria, 0.342 μm^3, is large compared to other published data. Salonen (1977) reported mean volumes ranging from 0.041–0.241 μm^3 in 12 Finnish lakes. Volumes reported by Coveney *et al.* (1977), Rheinheimer (1977), Traaen (1978), Stråskrabová & Komárková (1979), Jordan & Likens (1980) and Salonen (1981) fall within this range. However, meaningful comparisons are difficult to make because of the use of different methods and correction factors for shrinkage. Most similar to my data are those reported by Johansson (1981) which range from ~0.1–0.35 μm^3 for two acid (pH 5.0–5.5) forest lakes. The significance of the predominating large rod-shaped cells in Lake Gårdsjön is not clear.

Numbers of bacteria found in Lake Gårdsjön (0.6 $\cdot 10^{12}$–3.6 $\cdot 10^{12} \cdot m^{-3}$) were within ranges earlier recorded in oligotrophic water (Jordan & Likens

1980; Salonen 1981). Compared to Traaen's data (1978) they seem to be in the high range for acid clear water lakes. The high numbers occasionally found in the inlet stream showed that flooding of swampy areas along the inlet stream might resuspend substantial amounts of bacteria. As mentioned above this might also influence numbers of lake bacterioplankton in the epilimnion.

Relationships between phytoplankton and bacteria

The seasonal fluctuation of bacterioplankton biomass was partially correlated to the development of the total phytoplankton population. A positive relationship between bacteria and phytoplankton has been shown by Coveney *et al.* (1977), Silvey & Wyatt (1977) and Stråskrabová & Komárková (1979), and a negative relationship has been recorded by Oláh (1973). The exact course of development for the bacterial population compared to phytoplankton development could indicate a certain coupling between the two populations. Simultaneously increasing populations could indicate that bacterial consumption of released extra-cellular products is important (cf. Bell *et al.* 1974; Saunders 1975, 1977; Larsson & Hagström 1979). On the other hand, a delayed response could then indicate that autolysis products and dead phytoplankton cells are more important as found by Coveney *et al.* (1977). The difficulties in resolving the exact nature of the interrelationships, however, have been pointed out by Saunders (1977) and Hobbie & Rublee (1977). Thus the short-term diurnal changes studied and modelled by Saunders (1977) reveal a high degree of complexity including both abiotic and biotic factors.

The frequency of sampling in Lake Gårdsjön during the initial spring development in 1980 was obviously not high enough to give any details. The total phytoplankton biomass increased throughout April and reached a peak in May–June. Primary productivity during this period was also high (Grahn & Lydén, in prep.). The overall increase of bacterial biomass during April–May thus coincided with the increase in biomass and primary productivity. At this time, the biomass of bacteria was substantially smaller than phytoplankton biomass, especially in the hypolimnion. Biomass changes for phytoplankton and bacteria also showed a striking similarity during the summer. Minima in July and August in both epilimnion and hypolimnion thus coincided with those for phytoplankton. The increase during late summer was simultaneous within the respective layers as was the peak in September.

At this occasion, primary productivity was the highest recorded (Grahn & Lydén, in prep.) and biomass of bacteria was equal to or larger (in hypolimnion) than phytoplankton biomass. The following decrease was again closely correlated. The comparatively large biomass during late fall apparently had no connection to phytoplankton. However, littoral areas may have influenced bacterioplankton biomass at this time. In late fall degrading macrophytes and allochthonous leaf-litter may harbour large microbial populations as does superficial sediment. Through wave action and other water movements these populations can be resuspended in the water and may be of significance for the total bacterial population in the lake water.

From the above the importance of extra-cellular algal products compared to autolysis products and phytoplankton detrital matter for bacterioplankton during the summer of 1980 might be inferred. A final statement, however, can only be made based on much more detailed studies of bacterial activity and production during developmental phases of phytoplankton at a high degree of time resolution.

Acknowledgement

This work has been carried out within the integrated 'Lake Gårdsjön project' funded by the National Swedish Environment Protection Board (SNV), The Fishery Board of Sweden and the Swedish Council for Planning and Coordination of Research (FRN). Samples have been provided by O. Broberg and information on phytoplankton production and biomass has been provided by Olle Grahn. I would like to express my thanks to the field assistants S. Larsson and P. Leyton.

References

Bell, W. H., Lang, J. M. & Mitchell, R., 1974. Selective stimulation of marine bacteria by algal extracellular products. Limnol. Oceanogr. 19: 833–839.

Broberg, O. & Persson, G., 1981. Närsalttillgång i försurade sjöar. Fosfor, kväve och organiskt kol i Gårdsjön. Rep. Limnol. Inst., Univ. Upps. 55 pp.

Coveney, M. F., Cronberg, G., Enell, M., Larsson, K. & Olofsson, L., 1977. Phytoplankton, zooplankton and bacteria - standing crop and production relationships in an eutrophic lake. Oikos 29: 5-21.

Grahn, O., Hultberg, H. & Landner, L., 1974. Oligotrophication - a self-accelerating process in lakes subjected to excessive supply of acid substances. Ambio 3: 93-94.

Grahn, O. & Lydén, A., in prep. Primary producers and production in lake Gårdsjön.

Hobbie, J. E. & Rublee, P., 1977. Radioisotope studies of heterotrophic bacteria in aquatic ecosystems. In: Cairns, J. Jr (ed.). Aquatic Microbial Communities, pp. 441-476. Garland Publ. Inc., New York.

Hultberg, H., 1980. Metalleffekter i samband met försurning av sjöar. Meddn svenska Gruvföreningen 153(10): 3-24.

Johansson, J. Å., 1981. Development of bacterioplankton in small forest lakes. Proc. Symp. 28 September-2 October 1981, Färna, Skinnskatteberg, Sweden.

Jordan, M. J. & Likens, G. E., 1980. Measurement of planktonic bacterial production in an oligotrophic lake. Limnol. Oceanogr. 25: 719-732.

Larsson, U. & Hagström, Å., 1979. Phytoplankton exudate release as an energy source for the growth of pelagic bacteria. Mar. Biol. 52: 199-206.

Muniz, I. P., 1981. Acidification - effects on aquatic orgnaisms. In: Fazzolare, R. A. & Smith, C. B. (eds.). Beyond the Energy Crisis - Opportunity and Challenge, pp. 101-123. Pergamon Press, New York.

Oláh, J., 1973. Number, biomass, and production of planktonic bacteria in the shallow lake Balaton. Arch. Hydrobiol. 73: 193-217.

Overrein, L. N., Seip, H. M. & Tollan, A., 1980. Acid precipitation - effects on forest and fish. Res. Rep., Fr 19/80, SNSF-project, Oslo-Ås, 175 pp.

Rheinheimer, C., 1977. Microbiological investigations in rivers. II. Bacteria biomass of some river in North Germany. Arch. Hydrobiol. 81: 259-267.

Salonen, K., 1977. The estimation of bacterioplankton numbers and biomass by phase contrast microscopy. Ann. Bot. fenn. 14: 25-28.

Salonen, K., 1981. The ecosystem of the oligotrophic lake Pääjärvi. 2. Bacterioplankton. Verh. int. Ver. Limnol. 21: 448-453.

Saunders, G. W., 1975. Decomposition in freshwater. In: Anderson, J. M. & MacFadyen, A. (eds.). The Role of Terrestrial and Aquatic Organisms in Decomposition Processes, pp. 341-373. Blackwell Sci. Publ., Oxford.

Saunders, G. W., 1977. Carbon flow in the aquatic system. In: Cairns, J. Jr. (ed.). Aquatic Microbial Communities, pp. 417-440. Garland Publ. Inc., New York.

Silvey, J. K. G. & Wyatt, J. T., 1977. The interrelationship between freshwater bacteria, algae, and actinomycetes in southwestern reservoirs. In: Cairns, J. Jr. (ed.). Aquatic Microbial Communities, pp. 161-203. Garland Publ. Inc., New York.

Stråskrabová, V. & Komárková, J., 1979. Seasonal changes of bacterioplankton in a reservoir related to algae. I Number and biomass. Int. Revue ges. Hydrobiol. 64: 285-302.

Thompson, F. R. & Croll, N. E., 1974. Microbiology Report on Sudbury Environmental Study. Ont. Min. Envir. Tech. Rep. 39 pp.

Traaen, T. S., 1978. Bakterieplankton i innsjøer. TN 41/78, SNSF-project, Oslo-Ås. 16 pp.

Traaen, T. S., 1980. Effects of acidity on decomposition of organic matter in aquatic environments. In: Drabløs, D. & Tollan, A. (eds.). Ecological Impact of Acid Precipitation, pp. 340-341. SNSF-project, Oslo-Ås.

Zimmermann, R., Iturriaga, R. & Becker-Birck, J., 1978. Simultaneous determination of the total number of aquatic bacteria and the number thereof involved in respiration. Appl. envir. Microbiol. 36: 926-935.

Respiration of plankton in two small, polyhumic lakes

K. Salonen, K. Kononen & L. Arvola
Lammi Biological Station, University of Helsinki, SF-16900 Lammi, Finland

Keywords: forest water ecosystems, respiration, decomposition, mineralization, plankton

Abstract

The respiration of plankton of two polyhumic lakes was measured as production of carbon dioxide in dark bottles. The method proved to be enough sensitive for use in oligotrophic lakes with low alkalinity.

The respiration of plankton followed broadly changes in temperature. However, the primary production of phytoplankton was probably the main factor governing the seasonal pattern of respiration. During summer the respiration of plankton was more than three times higher than the primary production of phytoplankton. This suggests that allochthonous humic substances are an important source of carbon and energy for organisms of polyhumic lakes.

Introduction

Methodological difficulties have restricted knowledge of respiration rates of plankton in oligotrophic lakes (cf. Sorokin & Kadota 1972). The determination of oxygen has traditionally been used to estimate respiration as well as production of plankton. However even the more recent modifications of the Winkler procedure (Tschumi *et al.* 1978; Väätänen 1979) are not yet satisfactory to study respiration of plankton in oligotrophic, polyhumic lakes. In this study we have measured the respiration of plankton as carbon dioxide production. This approach has some theoretical advantages (Salonen & Holopainen 1979). When respiration is measured as carbon (carbon dioxide), no conversion factors to convert oxygen values to carbon are necessary. Overall, there are considerable advantages in measuring both production and destruction processes as well as biomass of organisms in organic carbon. The determination of dissolved inorganic carbon (Salonen 1981) is so simple and rapid that it is possible to make many more replicate determinations than with oxygen titrations. Because the relative sensitivities of oxygen and dissolved inorganic carbon determinations are about the same, it is advantageous that during summer in the epilimnion of weakly buffered forest lakes the relative change due to respiration is considerably higher in carbon than in oxygen. Moreover, such small lakes often have very thin oxygenated layers and the use of the oxygen method is not possible at all in the anoxic zone.

Study sites

The study lakes are the small lakes Horkkajärvi (0.01 km^2) and Nimetön (0.004 km^2) in southern Finland. The catchment areas of the lakes consist of forest at different stages of development. During summer the total nutrient concentrations in the uppermost 2–3 m varied between 500–700 mg N m^{-3} and 20–30 mg P m^{-3} in Lake Horkkajärvi and between 300–700 mg N m^{-3} and 10–40 mg P m^{-3} in Lake Nimetön. High concentrations of dissolved humic compounds characterize both lakes. The colour of water varied between 200–300 g Pt ·

Hydrobiologia 101, 65–70 (1983).

m^{-3} in Lake Horkkajärvi and 120–200 g Pt · m^{-3} in Lake Nimetön. The range of acidity was pH 5.5–6.0 in Lake Horkkajärvi. During the summer of 1980 the epilimnion of Lake Nimetön was acidified and this resulted in about 0.5 pH unit change in acidity which then varied between 4.9–5.8 in the uppermost one meter. More details of the biology of the lakes can be found in Arvola (this volume) and Rask (this volume).

Methods

Water samples were taken with a Ruttner-type sampler down to a depth of 2 m which includes the epilimnion and the upper part of the metalimnion. Samples were mixed thoroughly in a separate dark plastic vessel by turning up and down so gently that no air bubbles were introduced into the sample. Immediately after mixing two subsamples were siphoned into 65 ml borosilicate glass bottles which had been ignited (at 500 C) to remove any organic film on their surface and the residues of detergents. One of the bottles was transferred to the laboratory in ice and darkness and was processed there within about four hours of sampling. Dissolved inorganic carbon was determined according to Salonen (1981) in at least five replicates. After 24 h the other bottle incubated in darkness *in situ* at the depth of sampling was treated similarly. Respiration of plankton was calculated as a difference between final and initial concentrations of dissolved inorganic carbon. Primary production of phytoplankton was determined by radiocarbon method using GM detection of radioactivity.

Results

Because of their dark colour the study lakes had only a very thin warm epilimnion in summer. The concentration of oxygen decreased vertically with decreasing temperature until complete deoxygenation between 2–4 m depth (Fig. 1). The oxygen deficit in the hypolimnion results mainly from the meromictic nature of the lakes. When water temperature was more than 15 °C the respiration of plankton varied generally between 100–200 mg C · m^{-3} · day^{-1} at the surface of Lakes Horkkajärvi and Nimetön (Figs. 2, 3). With a few exceptions the highest rates were found near the surface. In the latter half of July a distinct maximum of respiration was observed between 1–2 m in Lake Nimetön. At this same time the level of chlorophyll was high and oxygen concentration was increasing in the same depth zone. The results from some other dates also indicated high respiration in the metalimnion, but these may be due to higher absolute variation in determinations made with water from 2 m containing up to 10 times more dissolved inorganic carbon than water at the lake surface.

In general the respiration of plankton was highest during the warmest period of the year. However, in June and July, in spite of stable temperatures, respiration fluctuated considerably (Fig. 4). In Lake Horkkajärvi the high rates seemed to follow high primary production of phytoplankton with some time delay. The only exception was found in September, but even then the peak of respiration was preceded by a clear peak in chlorophyll *a*. In lake Nimetön spring sampling began too late to give complete seasonal changes but the results still suggest that respiration and production peaks are successive rather than concurrent events in time. This seems more evident when we take into account the different time courses of primary production and respiration in water layers above and below 1 m in July in Lake Nimetön. In the hypolimnion primary production and respiration maxima were observed at the same time, but then high primary production

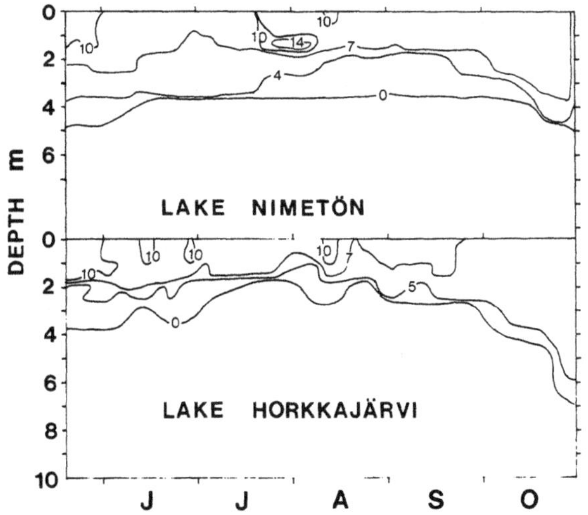

Fig. 1. Seasonal pattern of oxygen concentration in Lakes Horkkajärvi and Nimetön, 1980.

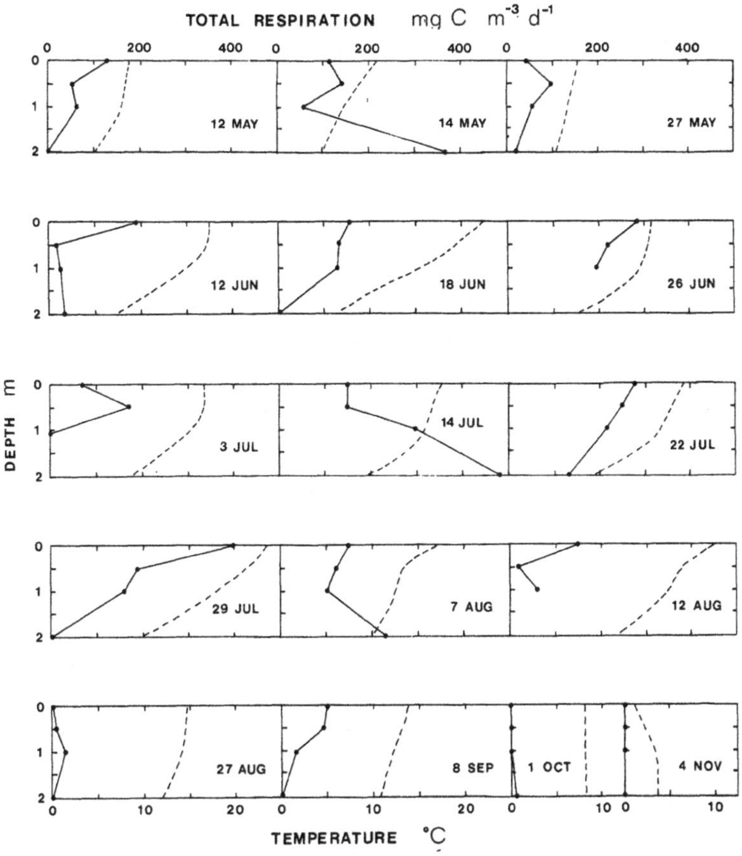

Fig. 2. Vertical distribution of temperature and respiration of plankton in Lake Horkkajärvi, 1980. Broken line – temperature; solid line – respiration of plankton.

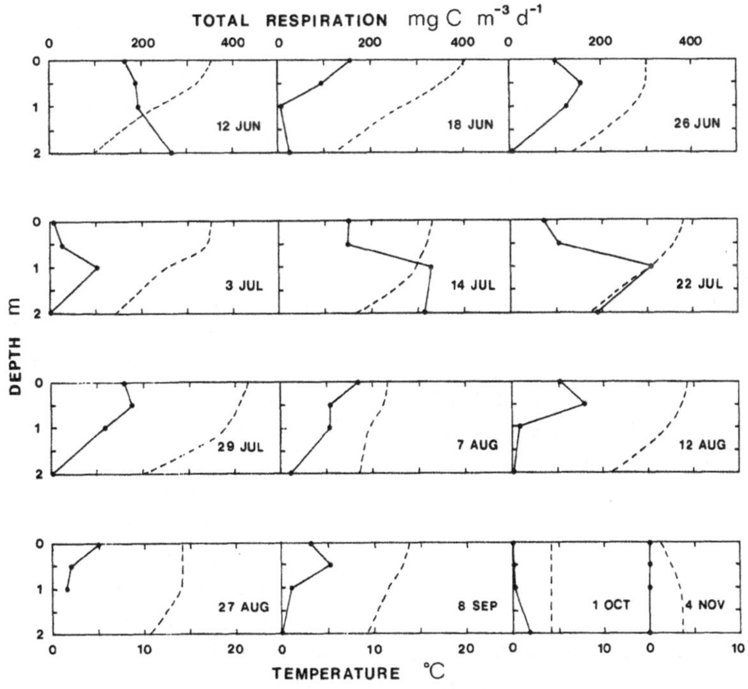

Fig. 3. Vertical distribution of temperature and respiration of plankton in Lake Nimetön, 1980. Broken line temperature; solid line respiration of plankton.

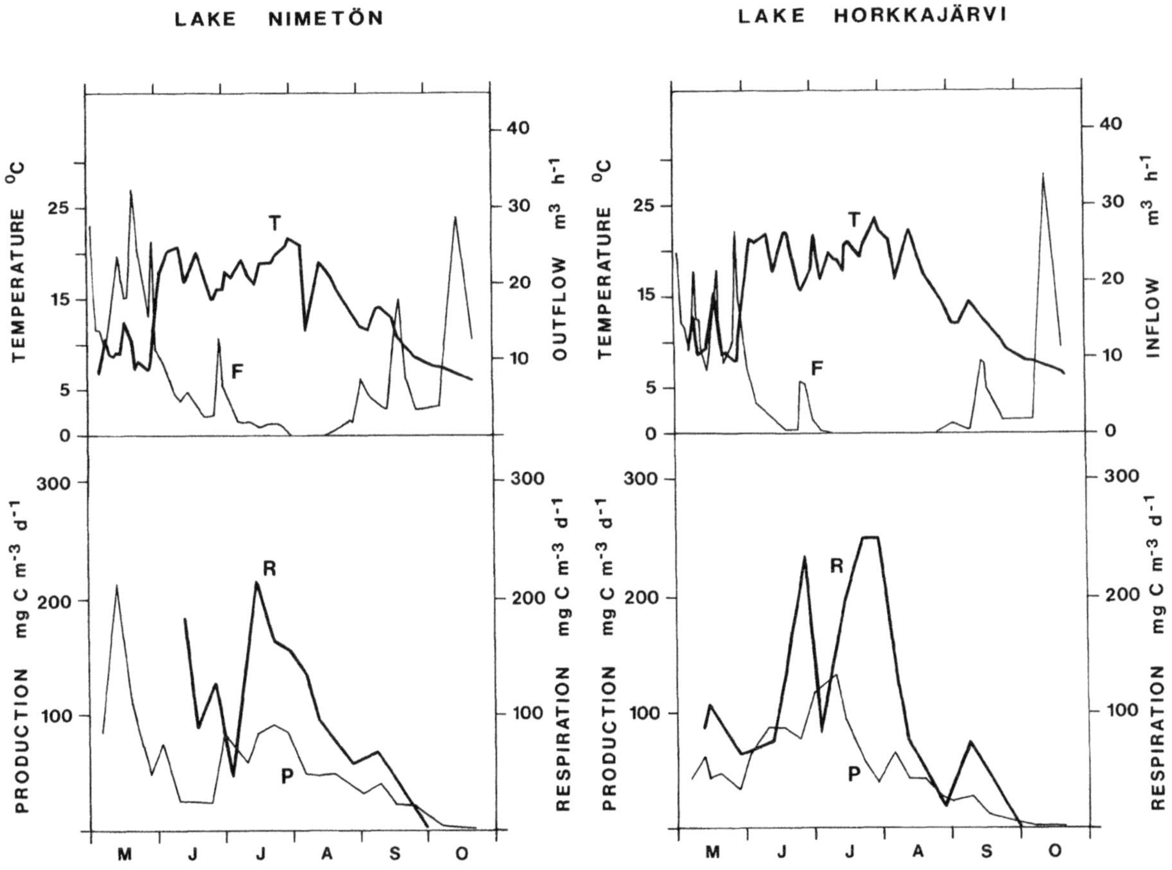

Fig. 4. Seasonal changes in surface temperature, flow of water, primary production of phytoplankton and respiration of plankton in Lakes Horkkajärvi and Nimetön, 1980.

below 1 m was preceded by high primary production near the surface. In autumn when primary productivity of phytoplankton decreased steadily and throughflow in the lakes increased, respiration of plankton was considerably lower than at equivalent temperatures in spring during higher primary productivity (Figs. 2, 4).

Between the beginning of June and the end of September plankton respired 19 g C · m^{-2} in Lake Horkkajärvi and 15.5 g C · m^{-2} in Lake Nimetön in the uppermost 2 m water column. These values are roughly three times higher than primary production of phytoplankton for the whole water column (Fig. 4).

Discussion

If we exclude possible effects of the enclosure of water samples in bottles, one of the main limitations in the determination of respiration by the dissolved inorganic carbon method is caused by changes in temperature. During the manipulation of samples before or after the experiment temperatures in bottles may increase so much that gas bubbles evolve. This may also happen in the lake during the incubation. These problems are particularly important in small, brown water lakes where water is usually oversaturated with carbon dioxide and where changes in water temperature at the surface can be very pronounced. Bubbling with air to stabilize carbon dioxide levels in samples is not advisable because the concentration of oxygen will usually change and bubbling may have some unknown effects on the organisms in the samples. In future when the determination of dissolved inorganic carbon is used to estimate primary production or respiration it would be advisable to manipulate samples in cooling boxes to avoid even slight warming. In this study five replicates were made for each determination of dissolved inorganic carbon. With 10 replicates the confidence limits of the results would be reduced by about 40%, but the time takan (4–7 min for all replicates) would still be acceptable. However, even the most precise results cannot necessarily be interpreted correctly if they are not representative. In our experience variation in time is particularly important in small forest lakes. We have found a case where a sampling interval of only 3 days would have been inadequate to detect a short pulse of primary production leading up to 10 mg C $\cdot$ l^{-1} of particulate organic matter compared with usual 1 mg C $\cdot$ l^{-1} level.

Our sampling for the determination of primary production and respiration was performed on different dates and therefore we were not able to use statistical methods to study the relationship between these results. Because the very shallow epilimnion of Lakes Horkkajärvi and Nimetön allows significant diurnal variation of water temperature, our single measurements of temperature are not adequate for precise analysis; a continuous record of temperature would be preferred. However, the measured surface temperatures probably give a good overall view of the seasonal changes in temperature of our experimental lakes. Broadly the respiration of plankton seemed to correlate with temperature. Considerable variation in respiration during stable temperatures in June and July and higher respiration in spring than in autumn suggest that temperature is not the most important factor governing respiration of plankton. Our results indicate that respiration increases shortly after the increase in primary production, but they do not reveal which organisms are responsible for the bulk of respiration. Along with our routine measurements we also fractionated plankton using 2 μm Nuclepore filters; this gave good separation of bacteria from other organisms (Salonen & Kononen, unpubl.). These results suggest that the bacterial component represents most of the respiration of plankton, although some anomalously high results force us to be cautious in stressing the importance of bacteria. The considerably higher respiration than primary production in the studied lakes supports the importance of bacteria.

In lakes Horkkajärvi and Nimetön the respiration of plankton was nearly three times higher than primary production of phytoplankton measured by the radiocarbon method. In fact the difference is higher when the entire seasonal cycle and whole lake down to the bottom are taken into account. The discrepancy between production and respiration may arise from inadequate methods. Although the radiocarbon method has been in use for about 30 years, it is still uncertain if the radiocarbon method in fact yields realistic estimates of primary production. Similar problems make the results of respiration also somewhat uncertain. However, the observed respiration and production values are possible, if allochthonous organic matter (about 20 g C $\cdot$ m^{-3} in both lakes) is an important source of carbon for organisms. The dependence of respiration on primary production might then be explained by co-metabolism (De Haan 1974) where substances derived from primary production function as a driving force for quantitatively larger decomposition of humic substances by bacteria.

Increased attention should be paid to the study of decomposition processes along with productive ones to better identify weaknesses in methods and to gain more reliable information on the roles of autochthonous and allochthonous production in the metabolism of lakes. We consider that our methodological approach is a useful tool in this respect, particularly in oligotrophic lakes with low alkalinity.

Acknowledgements

We thank Roger Jones for checking the language of our manuscript. This study was supported by the Maj and Tor Nessling Foundation and the Academy of Finland.

References

Arvola, L., 1983. Primary production and phytoplankton in two small, polyhumic forest lakes in southern Finland. Hydrobiologia (this volume).

Haan, H. De, 1974. Effect of a fulvic acid fraction on the growth of a Pseudomonas from Tjeukemeer (The Netherlands). Freshwat. Biol. 4: 301–310.

Rask, M., 1983. Differences in growth of perch (Perca fluviatilis L.) in two small forest lakes. Hydrobiologia (this volume).

Salonen, K., 1981. Rapid and precise determination of total inorganic carbon and some gases in aqueous solutions. Wat. Res. 15: 403–406.

Salonen, K. & Holopainen, A.-L., 1979. A comparison of methods for the estimation of phytoplankton primary production. Int. Revue ges. Hydrobiol. 64: 147–155.

Sorokin, Y. I. & Kadota, H., 1972. Techniques for the assessment of microbial production and decomposition in freshwaters. IBP Handbook 23, pp. 1–112.

Tschumi, P. A., Zbären, D. & Zbären, J., 1978. An improved oxygen method for measuring primary production in lakes. Verh. int. Ver. Limnol. 20: 43–48.

Väätänen, P., 1979. Microbial activity in brackish water determined as oxygen consumption. Arch. Hydrobiol. Beih. 12: 32–37.

Seasonal development of bacterioplankton in two forest lakes in central Sweden

Jan-Åke Johansson
Institute of Limnology, P.O. Box 557, S-751 22 Uppsala, Sweden

Keywords: forest water ecosystems, bacterioplankton, cell size, biomass, production, humic lakes

Abstract

Bacteria were counted with acridine orange epifluorescence technique in two humic lakes during 3 years. Less than 1% of the cells were found attached to detritus aggregates. 73% of the total number and 48% of the total volume were smaller than 1 μm. The mean cell volume ranged from 0.10 to 0.35 μm^3 with the highest cell volumes occurring during early summer contemporarily with the growth of the bacterial biomass and probably indicating favourable growth conditions. The mean density of bacteria in oligotrophic brown-water lakes is higher than in oligotrophic clear-water lakes. The development of bacterial biomass showed a regular and seasonally dependent pattern with maxima during early summer and autumn. The importance of different factors for the regulation of bacterial biomass is discussed. Three different approaches were used to estimate bacterial production. These resulted in an average production rate of 15–60 $\mu g\ C \cdot l^{-1} \cdot d^{-1}$ during the growing season. It was concluded that allochthonous sources comprised a significnat part of the energy supply to the bacteria in the two humic lakes.

Introduction

The seasonal development of planktic bacteria in humic lakes is poorly known. It was recently shown that variations in the size distribution of bacteria reflect their growth conditions (Larsson & Hagström 1982; Krambeck *et al.* 1981). An increased knowledge of the temporal variation in bacterial biomass and size composition thus increase the possibility to understand the role of planktic bacteria.

Investigations in humic lakes show that the autochthonous primary production alone cannot support the production of bacteria (Salonen 1981 and in prep.; Winberg 1982), suggesting that also allochthonous matter, mostly composed of humic compounds, should have constituted a significant energy source. These compounds are generally considered to be relatively resistant to microbial decomposition (Wetzel 1975). However, experiments have shown that humus may undergo biodegradation during certain conditions (De Haan 1972, 1975, 1977; Ryhänen 1968; Sederholm *et al.* 1973; A. Wilander, pers. comm.).

Results obtained during ecosystem studies in two small humic lakes (Eriksson *et al.* 1974, unpubl. results) initiated further studies on planktic bacteria. The aim of the present study was to elucidate seasonal changes in biomass and size structure of planktic bacteria in these two forest lakes and to make an assessment of the magnitude of the bacterial production.

The area and the lakes

Lake Botjärn and Lake Vitalampa are located some kilometers outside the village of Kloten (59° 54′N, 15° 50′E) about 200 km west of Uppsala (Fig. 1). The Kloten area is situated on the Scandin-

Hydrobiologia 101, 71–88 (1983).

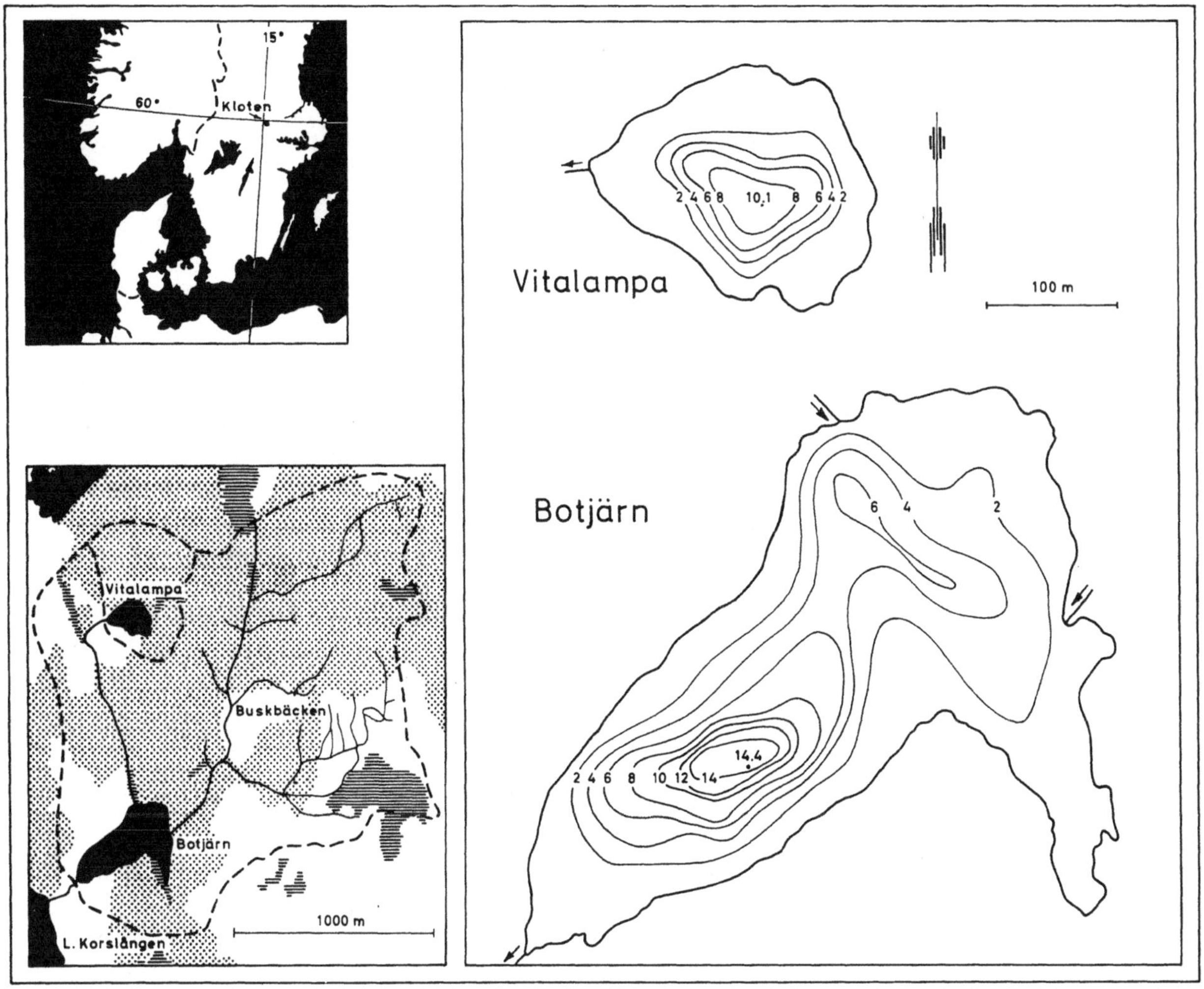

Fig. 1. Lake Vitalampa and Lake Botjärn. Drainage area. Vegetation in the catchment: young forest (<20 years old □), older forest (>20 years old ▩), marsh (▤). Bathymetric maps.

Table 1. Meteorological data of the Kloten area. Mean 1970–1977 (after Grip 1982).

Precipitation (mm · $year^{-1}$	927
Runoff (mm · $year^{-1}$	443
Temperature, mean for the year (C)	4.9
Temperature, mean for January (C)	-3.7
Temperature, mean for July (C)	15.3
Duration of snow-cover (days)*	140
Length of growing season (days)*	185

* 1901–1930 (Atlas över Sverige, Stockholm, 1953).

Table 2. Yearly mean and range of some chemical data for the period 1975–1976.

	Lake Botjärn		Lake Vitalampa	
	$\bar{x}$	range	$\bar{x}$	range
Total N (mg · l^{-1})	0.37	0.19–0.50	0.33	0.09–0.50
Total P (mg · l^{-1})	0.018	0.007–0.028	0.014	0.009–0.031
pH	5.4	4.7–6.3	5.1	4.4–5.3
Colour (mg Pt · l^{-1})	57	15–100	41	10–70

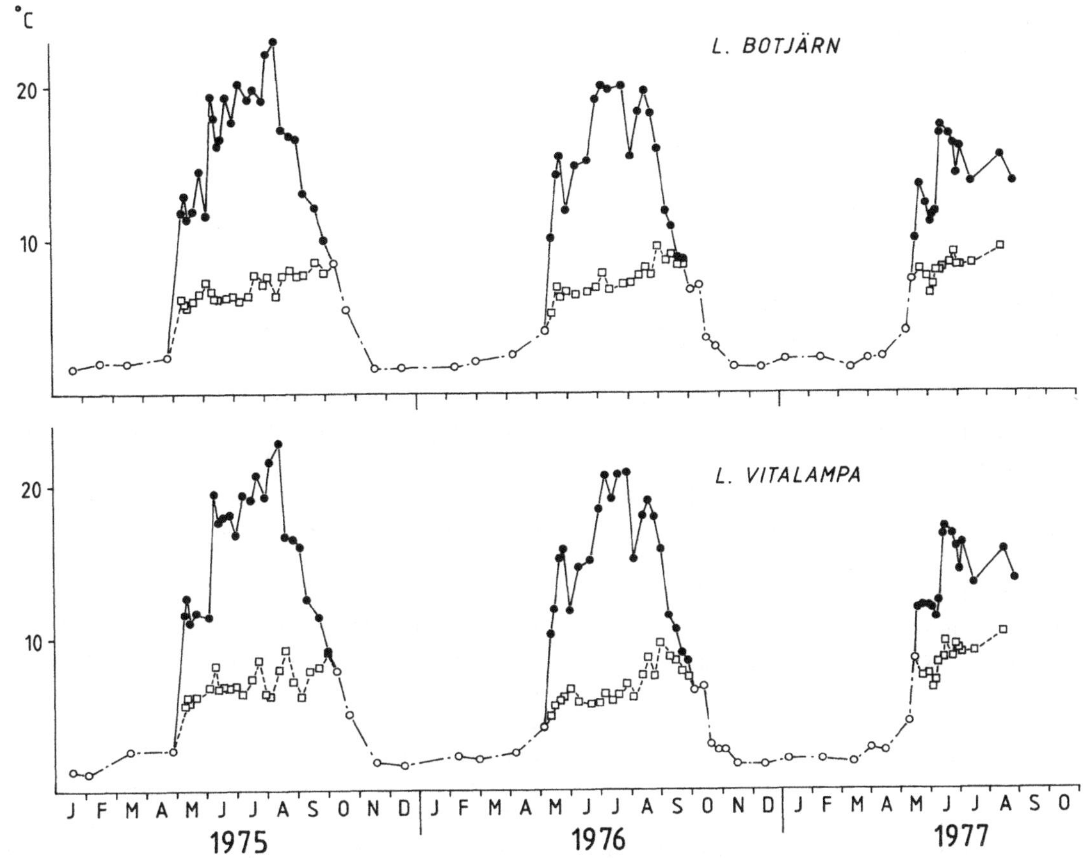

Fig. 2. Temperature in the epilimnion (●), the hypolimnion (□) and the whole water volume (○) of Lake Botjärn and Lake Vitalampa.

avian table land in the northern boreal coniferous region. The climate is rather severe, with long winters and rather high precipitation (Table 1). Lake Vitalampa is a seepage lake draining into a small stream which empties into Lake Botjärn 1 km downstream (Fig. 1). The surfaces of the lakes are 9.7 and 2.9 ha and the mean depths 3.3 and 2.8 m in Lake Botjärn and Lake Vitalampa, respectively. The theoretical turnover time is 0.3 year for Lake Botjärn and 1.5 years for Lake Vitalampa.

The lakes are poor in nutrients and weakly acid (Table 2). The colour is about 60 mg Pt $\cdot$ l^{-1} in Lake Botjärn and 40 mg Pt $\cdot$ l^{-1} in Lake Vitalampa. The concentrations of DOC are about 10 and 8 mg $\cdot$ l^{-1} in Lake Botjärn and Lake Vitalampa, respectively. A thermocline, at 2–4 m depth, has usually developed within a week after the ice has disappeared from the lakes, which occurred during the last days of April in 1975 and one week later in 1976 and 1977.

Methods

Sampling

Water samples were collected weekly (May–October) and monthly (November–April) at the point of maximal depth of the lakes. Before March 1977 the samples (excluding those of phytoplankton) were taken with a Ruttner sampler from the depths 1, 3, 6 m and also from 9 m in Lake Vitalampa and 12 m in Lake Botjärn. After that time, integrated samples were collected with a 2 m tube (Ramberg 1976) from 0–4 m in both of the lakes and from 4–8 m in Lake Vitalampa and from 4–10 m in Lake Botjärn. Integrated phytoplankton samples were taken from 0–3 m with a Ruttner sampler.

All data, except those obtained from integrated sampling, are weighed by water volume within the respective epilimnion and hypolimnion or within the total water body when the lakes were not stratified.

Environmental factors

After removal of zooplankton and larger particles with a 40 μm net, water was filtered through a pre-ignited Whatman GF/F-filter and the particulate organic carbon content (POC) measured with a LECO WR-12 carbon analyzer. Samples to be analyzed for dissolved inorganic carbon (DIC) were preserved with $HgCl_2$ in 100 ml glass bottles and kept cool before they were analyzed with a gas chromatograph according to Stainton (1973). Monthly samples were taken from the surface to the bottom at 2 m intervals and analyzed for dissolved oxygen with Winkler technique.

Phytoplankton counting and biomass calculations are described by Blomqvist *et al.* (1981). Phytoplankton production was measured with ^{14}C-technique in 120 ml glass bottles exposed *in situ* for 24 h at the depths 0, 0.25, 0.5, 1, 2, 3 and 4 or 5 m. Determination of radioactivity was identical with the method used in the dark uptake measurements (see below).

Biomass of bacterioplankton

Bacteria were counted with a modified method of the acridine orange direct-count technique (AODC) described by Zimmermann *et al.* (1974). The samples were preserved in 100 ml glass bottles provided with formalin to a final concentration of 2%. A 5–10 ml subsample was mixed with a corresponding volume of newly prepared acridine orange solution in a glass funnel equipped with a 0.2 μm Nuclepore polycarbonate filter (47 mm) placed on a sintered glass plate. After 2 min staining the solution was filtered at a vacuum pressure of 0.15 atm and rinsed twice with 3.5 ml water to reduce background fluorescence, after which the filter was dried at room temperature. pH was about 7.8 during the staining procedure. All solutions were prefiltered before use (0.2 μm).

Two conditions had to be fulfilled in order to minimize an otherwise very heterogeneous distribution of the cells on the filter surface: first, the sintered glass plate was pre-wetted and the filter floating on the water film was brought into contact with the plate by aid of vacuum pressure; secondly, during the filtration procedure, the rinsing water had to be added before the meniscus of the solution reached the filter surface.

Table 3. Size (μm) and volume (μm^3) of cells.

	Cocci	Rods	Rods or others			
Length group	<1	<1	1–2	2–3	3–5	>5
Length	0.6	0.6	1.5	2.5	4.0[a]	[b]
Width	0.6	0.5	0.5	0.6	0.6[a]	[b]
Volume	0.113	0.113	0.295	0.707	1.131[a]	[b]

[a] Usually measured.
[b] Always measured.

The counting was made with a Zeiss standard RA microscope equipped with an illuminator 100 for epifluorescence. A 50 W super pressure mercury lamp was used together with excitation filters KP 490 and KP 500, a barrier filter LP 528 and a reflector Rfl 510. A 100 × neofluar objective was used together with 12.5 × KPL ocular equipped with a square (25 × 25 μm) and a micrometer scale.

Depending on the density of cells on the filter, between 15 and 30 randomly chosen square fields were counted (on average 400 cells in total). The counting accuracy was ±5% calculated as the 95% confidence limits of the total number.

For calculation of the volumes, about 40% of the counted cells were grouped into 5 size groups according to length and the cell volumes were calculated according to Table 3. The width used was based on measurements from about 400 cells. Both width and length were usually measured on cells longer than 3 μm and always on cells longer than 5 μm. Cylinder shape was assumed when the volume of rod-shaped cells (rods and vibrions) were calculated. The carbon content of bacteria was assumed to be 10% by volume (Sorokin & Kadota 1972).

Bacterial production

Three different approaches were used to estimate bacterial production:

1. The dark CO_2 uptake was measured according to Kuznetsov & Romanenko (1966) and Overbeck (1979). To 120 ml glass-stoppered dark bottles were added 1 ml $NaH^{14}CO_3$ (1 μC, 0.35 μmol). The bottles were exposed *in situ* for 24 hours at the depths 1, 3, 6 and also at 9 m in Lake Vitalampa and 12 m in lake Botjärn. After filtration through 0.2 μm filters (SM 11107) and rinsing with distilled water

the filters were placed in Soluene TM-350 (Packard) for 24 hours at room temperature. The ^{14}C activity was measured in a liquid scintillation spectrometer (Intertechnique SL-30) in a toluene-based scintillation cocktail (50 mg PPO and 5 mg POPOP per 10 ml toluene). Counting efficiency was determined with internal standard method (Hexadecane).

In the epilimnion it was assumed that the dark uptake of CO_2-carbon comprised 6% of the total bacterial production, P (op. cit.), i.e., $P = \frac{100 \times A}{6}$ ($\mu g\,C \cdot l^{-1} \cdot d^{-1}$) where A = dark CO_2 uptake ($\mu g\,C \cdot l^{-1} \cdot d^{-1}$). The production estimates obtained by using a constant value of 6% must be regarded as approximative as it has been shown that this percentage share may vary widely (Overbeck 1979).

In the hypolimnion of the lakes low oxygen concentrations and the contact with reduced sediments implied that the probability of the presence of heterotrophs using a considerable share of external CO_2-carbon in their biosynthesis and the presence of true chemoautotrophs was large (Sorokin 1965). The dark uptake in the hypolimnion thus represented a measure of the chemoautotrophic bacterial production plus that share of the heterotrophic bacterial production which was based on the CO_2-carbon uptake. For simplicity these two groups of bacteria are called chemosynthetizers (op. cit.).

2. During June–August 1975, the community respiration of CO_2 was measured as the difference between initial and final concentrations of DIC in 120 ml dark bottles exposed for 24 hours at a depth of 1 m in the lakes. The bacterial part of the respiration was calculated in the following way: the contribution from zooplankton to the total respiration was estimated to be 10%, based on the average zooplankton biomasses during the period in question and respiration rates from Ivleva (1980). Phytoplankton constituted, on average, 12% and 20% of the total volume of living seston (excl. zooplankton) in Lake Vitalampa and Lake Botjärn, respectively. Assuming that respiration from phytoplankton was proportional to its share of the total biomass a rough estimate of the bacterial part of the respiration was obtained by subtracting the calculated respiration of zoo- and phytoplankton from the total production of DIC.

Production estimates from respiration measurements depend on the growth efficiencies (net production/gross production). Reported values vary widely, but a growth efficiency of 0.4 (Jordan & Likens 1980; Winberg 1980) was assumed to be a reasonable value.

3. The apparent rates of DIC accumulation and O_2 consumption were calculated as the differences between the highest and lowest concentrations below the thermocline during May–August 1975 and 1976 and below 4 m when the lakes were ice-covered, as the through-flow of water near the ice was assumed to be more intense implying that losses of CO_2 could be severe. The total bacterial production was then calculated from the production of CO_2-carbon, assuming a growth efficiency of 0.4 as above. The aerobic part of this CO_2 production was calculated from the oxygen consumption assuming that 1 mole O_2 corresponded to 1 mole CO_2-carbon.

Results

Total number

The contribution of cells attached to detritus aggregates was less than 1% of the total number. The low share depends on the sparse occurrence of such particles in the counts. In Lake Vitalampa detritus aggregates were observed in 9% of the samples from the depths of 1 m + 3 m, in 21% at 6 m depth and in 27% of those from 9 m. The frequency in Lake Botjärn was 13% at 1 m + 3 m and 25% at 6 m and 12 m, respectively. The aggregates were, however, densely colonized by cells. Thus the average increase in cell number was almost 70% in square fields where detritus aggregates were present. No bacteria were observed on the few phytoplankton cells occurring in the counts.

Total number of bacteria ranged from $1.2\text{–}7.3 \times 10^6$ cells ml^{-1} in Lake Botjärn and from $0.7\text{–}8.2 \times 10^6\ ml^{-1}$ in Lake Vitalampa. The differences between depths were sometimes large, especially during late summer when high densities occurred in the hypolimnion. In Lake Vitalampa the mean number was $3.7 \times 10^6\ ml^{-1}$ at all sampling depths except at 9 m where the density was $4.8 \times 10^6\ ml^{-1}$ (Table 5). In Lake Botjärn the mean density was $3.1 \times 10^6\ ml^{-1}$ at 1 m and increased somewhat with depth to $3.4 \times 10^6\ ml^{-1}$ at 12 m. The seasonal variation in total number resembles that of the total volume treated below.

Table 4. The relative size distribution of planktic bacteria. Average for all samples of the lakes.

Length class (μm)	<1	1-2	2-3	3-5	>5
Total number (%)	73.1	23.1	2.4	1.0	0.4
Total biomass (%)	48.0	36.0	8.6	5.3	2.1

Cell size

The most common morphological cell types were single or paired rods and vibrions, which constituted the greatest share of the biomass. However, many of the smallest cells counted, i.e. shorter than 0.5 μm, were difficult to distinguish as regards shape (cocci, rods, etc.), because their size is near the limit of resolution of the microscope. There is also a risk that cells have settled obliquely on the filter surface and thereby exposed a distorted picture of their shape.

The average size distribution in different length classes is illustrated in Table 4. 73% of the total number and half of the total volume belonged to cells shorter than 1 μm. Cells longer than 3 μm, corresponding to cell volumes larger than 1 μm^3, comprised only about 1% of number and 7% of volume.

The size distribution differed between depths (Table 5). At 1 m, and also at 9 m in Lake Vitalampa, the share of the smallest size category (<1 μm) was, on average, 72% of the total number. However, at 6 m the percentage was significantly higher ($p < 0.01$) in both lakes.

The mean volume per cell ranged from 0.10–0.35 μm^3 during the investigation period. It was usually low in winter, on average 0.17 μm^3 in Lake Vitalampa and 0.18 μm^3 in Lake Botjärn during October–April. In May, from ice-break on and during the period of warming-up and development of a thermocline, there was a sharp increase in 1975 and 1976 (Fig. 3). The mean volume at these maxima was between 0.25 and 0.32 μm^3. In Lake Botjärn the maxima occurred in late May or early June, in Lake Vitalampa 1–2 weeks earlier. They coincided with a contemporary increase in cell density of the bacteria, resulting in a pronounced increase in total volume as well (see next section). In the hypolimnion there was a corresponding but less pronounced and somewhat delayed increase in mean volume.

In congruence with the vertical differences in size distribution (see above), the smallest mean volumes were found at 6 m and the largest in the surface and bottom waters (Table 5). The differences between 1 m and 6 m were significant in both lakes ($p <$

Table 5. Share of cells shorter than 1 μm, total cell number, mean cell volume and total biomass of bacteria during the period June 1975–February 1977. Mean of all samples ($\bar{x}$), coefficient of variation (cv) and number of sampling occasions when registration was made (n).

	Lake Botjärn				*Lake Vitalampa*			
	Depth (m)	$\bar{x}$	cv	n	Depth (m)	$\bar{x}$	cv	n
<1 μm (% of total number)	1	72	9	44	1	72	13	41
	3	74	9	44	3	73	14	35
	6	76	10	44	6	77	13	41
	12	74	9	44	9	72	10	41
Total number (10^6 ml^{-1}	1	3.10	35	50	1	3.70	38	56
	3	3.19	34	50	3	3.70	38	56
	6	3.27	28	50	6	3.71	38	56
	12	3.43	28	50	9	4.81	38	56
Mean volume (μm^3)	1	0.203	17	44	1	0.186	18	41
	3	0.190	20	44	3	0.181	17	35
	6	0.183	18	44	6	0.173	16	41
	12	0.188	13	44	9	0.202	20	41
Total biomass ($\mu m^3 \cdot 10^6 \cdot ml^{-1}$)	1	612	31	44	1	686	37	41
	3	574	30	44	3	659	34	34
	6	599	36	44	6	654	42	41
	12	654	35	44	9	988	40	41

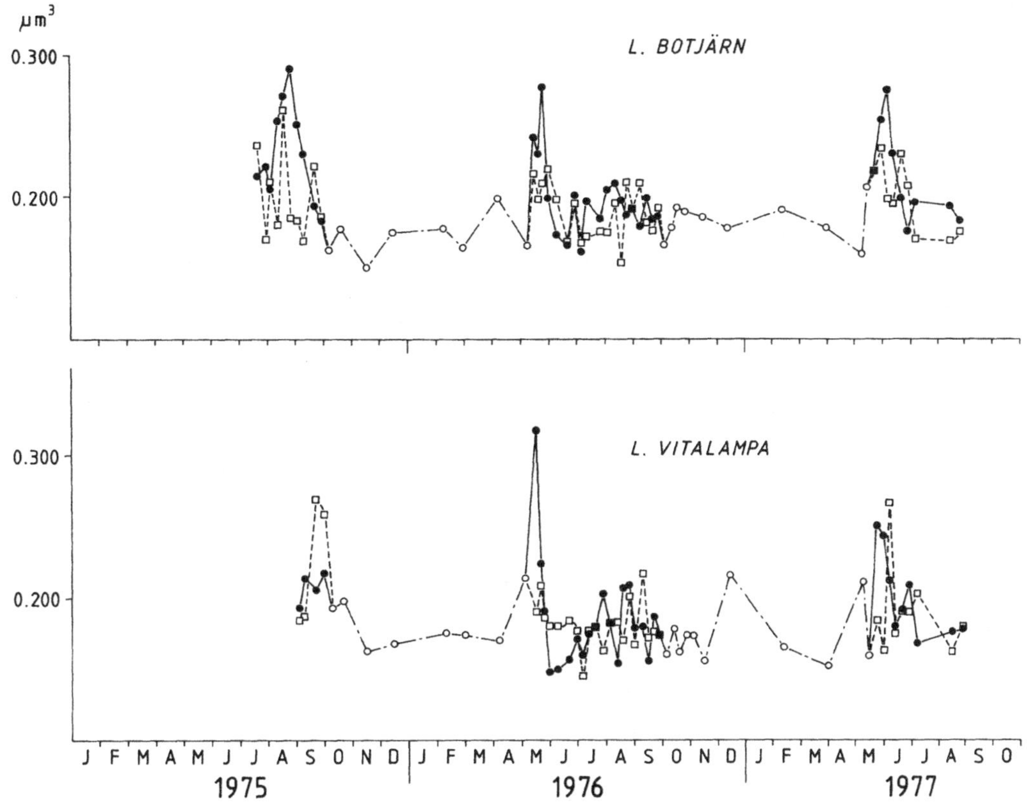

Fig. 3. Bacterial mean cell volume in the epilimnion (●), the hypolimnion (□) and the whole water volume (○) of Lake Botjärn and Lake Vitalampa.

0.05), and also those between 6 m and 9 m in Lake Vitalampa ($p < 0.001$).

Biomass

In both lakes the seasonal variation in total volume of bacteria was characterized by two maxima of considerable size and duration (Fig. 4). The first started to develop in connection with ice-break, forming a peak in May or June. The second maximum, which was of the same magnitude as that of early summer, occurred in September or October.

During the first week after ice-break the development in the hypolimnion was similar to the biomass changes in the epilimnion but differed later on during the course of the summer (Fig. 4). This was especially evident in Lake Vitalampa in 1976, when the total volume of bacteria in the hypolimnion remained very large throughout the summer.

Bacterial activity and calculations of production

The dark CO_2 uptake rates in the epilimnion were rather high at the first measurements in May 1975 and 1976 (Fig. 5). The highest uptake rates were, however, measured later during summer, except in Lake Vitalampa in 1975 when no increase occurred after the peak in May. The influence of temperature on the dark uptake was probably high in the epilimnion. When log-specific dark uptake (dark assimilation per bacterial volume) was plotted against the reciprocal of absolute temperature in the Arrhenius equation, the Q_{10} values were about 5 (r^2 0.82 and 0.75 in Lake Botjärn and Lake Vitalampa, respectively). The mean dark CO_2 uptake in the epilimnion was 1.5 $\mu g\ C \cdot l^{-1} \cdot d^{-1}$ and 1 $\mu g\ C \cdot l^{-1} \cdot d^{-1}$ in Lake Botjärn and Lake Vitalampa, respectively, corresponding to production rates of about 25 and 17 $\mu g\ C \cdot l^{-1} \cdot d^{-1}$ (Table 7).

The dark CO_2 uptake in the hypolimnion was higher than in the epilimnion and also differed considerably, both in magnitude and seasonal devel-

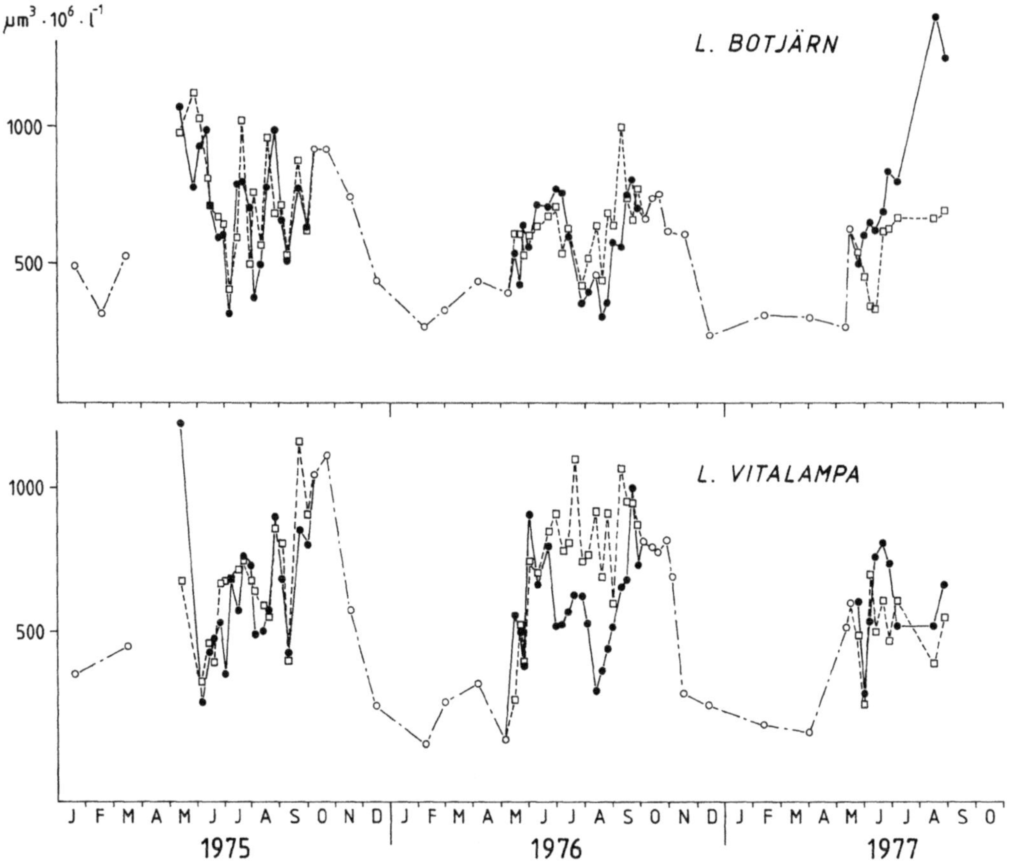

Fig. 4. Total bacterial volume in the epilimnion (●), the hypolimnion (□) and the whole water volume (○) of Lake Botjärn and Lake Vitalampa.

opment between the years (Fig. 6). During 1975 the highest rates were measured in late summer in both lakes. The mean values were 1.9 and 3.0 $\mu g\ C \cdot l^{-1} \cdot d^{-1}$ in Lake Botjärn and Lake Vitalampa, respectively (Table 6). During the following year the dark CO_2 uptake increased rapidly from June and remained very high until late August. In Lake Vitalampa the maximum value was 46 $\mu g\ C \cdot l^{-1} \cdot d^{-1}$ in the end of June, and during 1.5 months the dark uptake exceeded 15 $\mu g\ C \cdot l^{-1} \cdot d^{-1}$. The mean value for 1976 was 14.4 $\mu g\ C \cdot l^{-1} \cdot d^{-1}$ in this lake and 5.7 $\mu g\ C \cdot l^{-1} \cdot d^{-1}$ in Lake Botjärn. The dark uptake rates (Table 6) represent minimal estimates of the cheosynthetic bacterial production in Table 7.

The community respiration of CO_2 during June–August 1975 was, on average, 96 and 114 $\mu g\ C \cdot l^{-1} \cdot d^{-1}$ in Lake Botjärn and Lake Vitalampa, respectively. The range was from near zero to 216 $\mu g\ C \cdot l^{-1} \cdot d^{-1}$ and the coefficients of variation 66 in Lake Botjärn and 49 in Lake Vitalampa. Subtracting the estimated respiration of zooplankton and phytoplankton, the respiration of bacteria was calculated to 69 and 90 $\mu g\ C \cdot l^{-1} \cdot d^{-1}$ (Table 6) and the corresponding production rates to 46 and 60 $\mu g\ C \cdot l^{-1} \cdot d^{-1}$ in Lake Botjärn and Lake Vitalampa, respectively (Table 7).

The DIC content increased considerably during winter when the lakes were covered with ice and in summer below the thermocline (Fig. 7). At the end of these periods the partial pressure of CO_2 was 20–30 times higher than that of the atmosphere. As O_2 developed in a reverse pattern to that of CO_2, the O_2 concentration was very low at the end of the stagnation periods (Fig. 7). During overturn in autumn there was a rapid release of CO_2 to the atmosphere and an almost complete oxygenation of the whole water bodies. The usually short periods be-

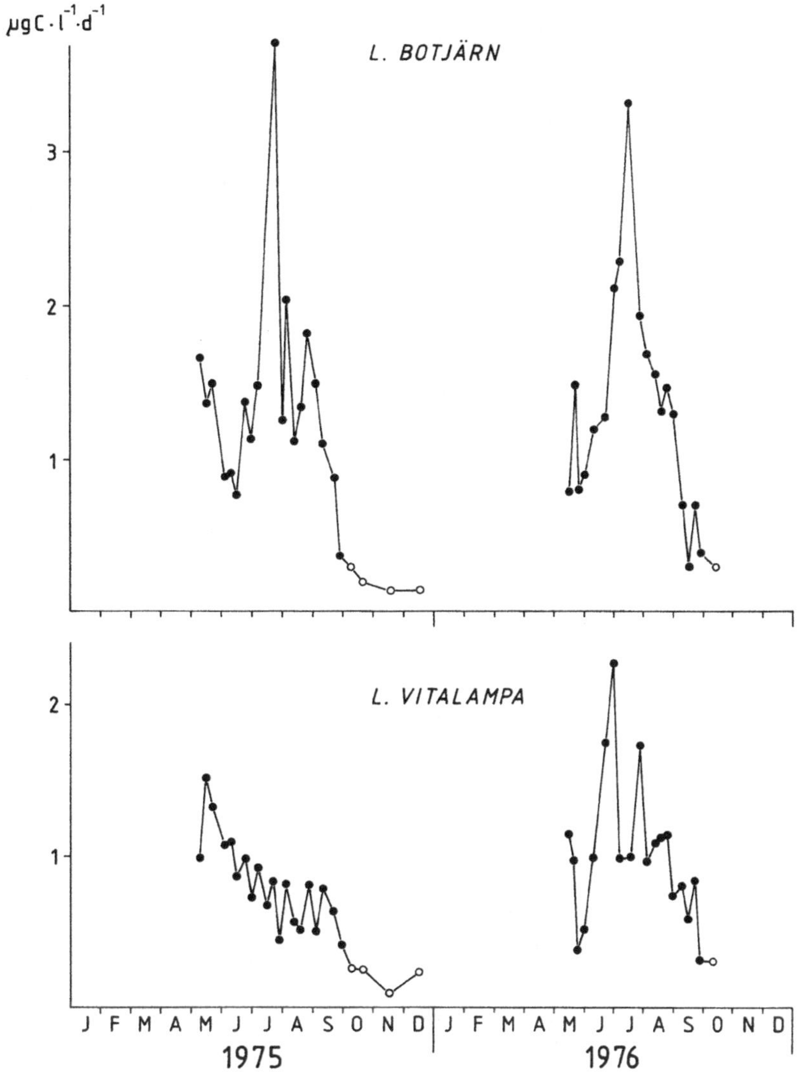

Fig. 5. Dark $^{14}CO_2$ uptake in the epilimnion (●) and the whole water volume (○) of Lake Botjärn and Lake Vitalampa.

tween ice-break and development of a thermocline caused an incomplete spring circulation and poor aeration of the deeper water masses. The increase in oxygen saturation at these depths was only 20% in Lake Botjärn and 30% in lake Vitalampa in 1975 and 5% and 8% during the following spring.

The respiration rates in the hypolimnion and below the depth of 4 m in winter (Table 6) imply that CO_2 production is higher than O_2 consumption both in winter and summer. The average CO_2 production:O_2 consumption ratios were between 1.7 and 2.9, indicating that about half of the hypolimnetic respiration was of anaerobic origin. The calculated production rates were between 15 and 35 μg C $\cdot\ l^{-1} \cdot d^{-1}$ (Table 7).

Discussion

Total number

The concentration of organic matter influences the density of bacteria, which is evident from Table 8 where the brown oligotrophic lakes have higher bacterial densities compared to the clear lakes. The quality of the organic matter is also important. The

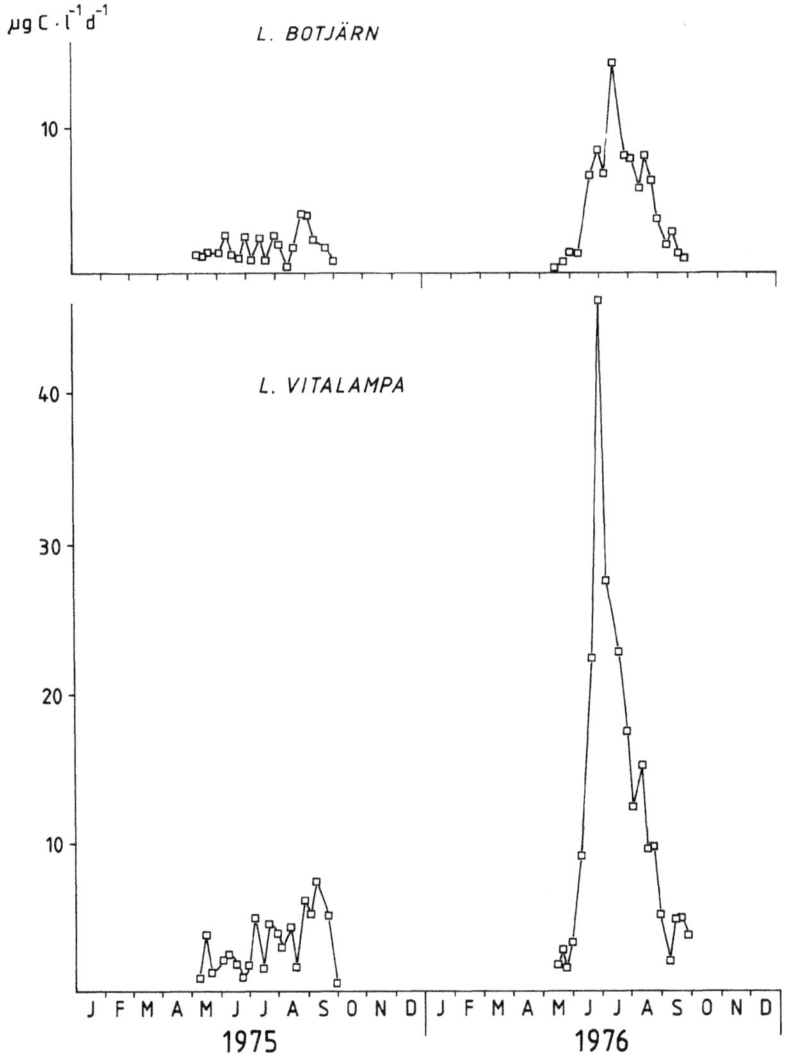

Fig. 6. Dark $^{14}CO_2$ uptake in the hypolimnion of Lake Botjärn and Lake Vitalampa.

Table 6. Bacterial activity ($\mu g\ C \cdot l^{-1} \cdot d^{-1}$) measured with different methods (see text).

	Lake Botjärn					Lake Vitalampa				
	Epilimnion		Hypolimnion		>4 m[b] 1975–1976	Epilimnion		Hypolimnion		>4 m[b] 1975–1976
Dark $^{14}CO_2$ uptake	1.5	1.6	1.9	5.7		0.9	1.1	3.0	14.4	
Dark bottle respiration[a]	69					90				
Total CO_2 accumulation			31	44	22			37	53	33
Aerobic CO_2 accumulation			15	16	13			19	23	17

[a] June–August, 1 m.
[b] Ice-covered period.

Table 7. Calculated bacterial production ($\mu g\ C \cdot l^{-1} \cdot d^{-1}$) calculated from data in Table 6 (see text).

	Lake Botjärn					Lake Vitalampa				
	Epilimnion		Hypolimnion		>4 m 1975–1976	Epilimnion		Hypolimnion		>4 m 1975–1976
Dark $^{14}CO_2$ uptake:										
a) heterotrophy	25	27				15	18			
b) chemosynthesis			2	6				3	14	
Dark bottle respiration	46					60				
Total CO_2 accumulation			21	29	15			24	35	22
Aerobic CO_2 accumulation			10	10	8			13	15	11

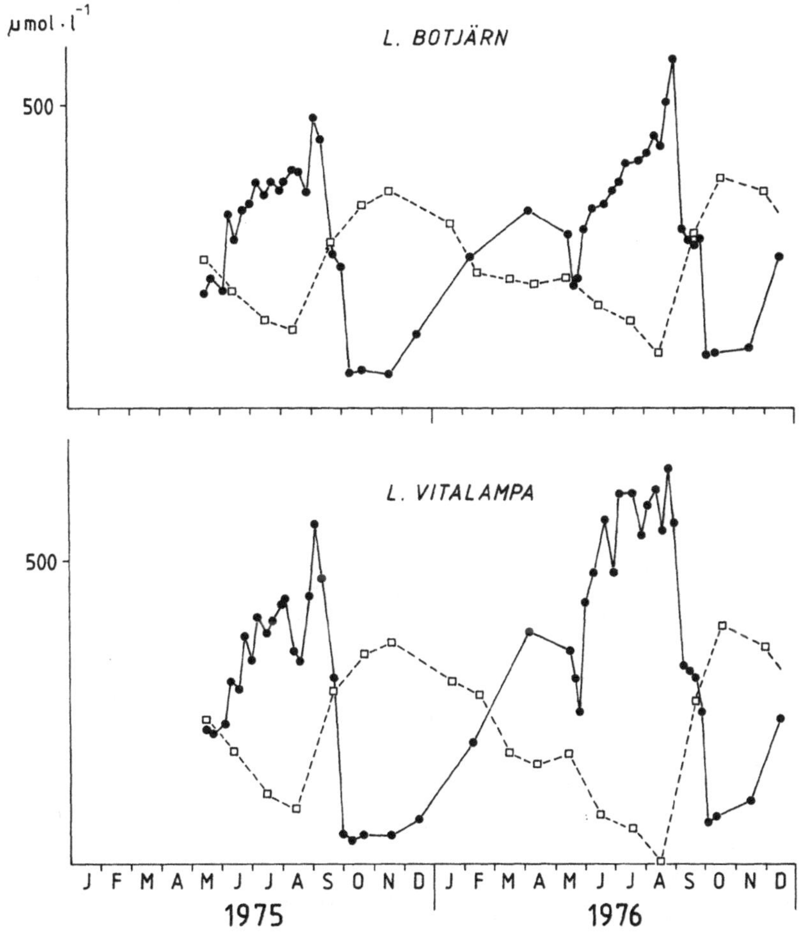

Fig. 7. Concentrations of dissolved inorganic carbon (●) and oxygen (□) in the hypolimnion during summer stratification and below 4 m during the rest of the year.

Table 8. Mean and range of total bacterial number ($10^6 \cdot ml^{-1}$) from surface or epilimnion waters during growing season.

Lake	Mean	Range	(Authors)	
Gunillajaure	0.02	0.01–0.03	(J. Johansson 1980)	
Stugsjön	0.13	0.05–0.3	(Persson 1977)	
Mesvattnet	0.8	0.5–1.1	(Johansson unpubl.)	Oligotrophic
Småvattnet	1.0	0.9–1.3	(Johansson unpubl.)	clear-water
Gårdsjön	~1.4	0.6–3.65	(Andersson in prep.)	lakes
Mirror lake[a]	1.6		(Jordan & Likens 1980)	
Tingvallavatn	2.2	1.4–3.3	(Johansson unpubl.)	
Fisklösen	2.5	1.0–4.6	(Johansson unpubl.)	
Vitalampa 1975	3.7	1.9–4.9	(This study)	
Vitalampa 1976	3.9	2.1–6.8	(This study)	
Vitalampa 1977	3.6	1.6–4.8	(This study)	Oligotrophic
Botjärn 1975	3.3	1.8–5.4	(This study)	brown-water
Botjärn 1976	3.0	1.6–4.7	(This study)	lakes
Botjärn 1977	3.7	1.7–7.3	(This study)	
Grästjärn	2.7	1.7–5.3	(Johansson unpubl.)	
Orrtjärn	3.3	1.4–5.9	(Johansson unpubl.)	
Norrviken	5.0	3.0–9.4	(I. Ahlgren unpubl.)	Eutrophic
Bysjön	~6.6	3.9–13	(Coveney et al. 1977)	lakes
Mossø[b]	~14	11–18	(Riemann et al. 1982)	

[a] Weighed yearly mean.
[b] March 24–May 5.

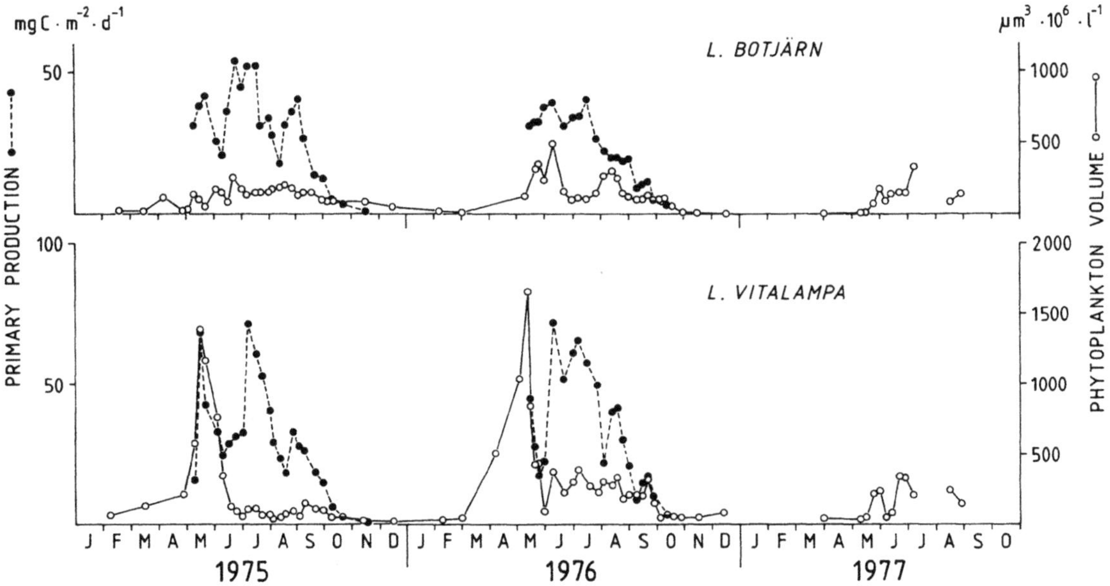

Fig. 8. Production and biomass of phytoplankton.

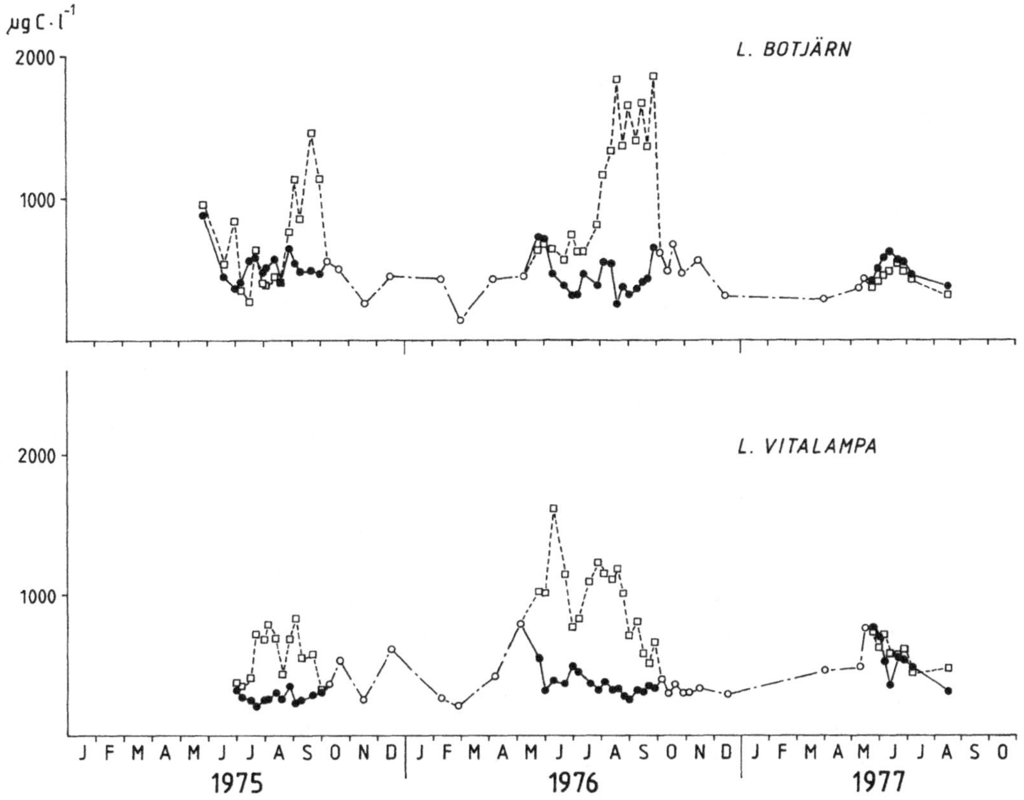

Fig. 9. Concentration of particulate organic carbon in the epilimnion (●), the hypolimnion (□) and the whole water volume (○) of Lake Botjärn and Lake Vitalampa.

concentration of organic carbon is approximately the same in the eutrophic and brown oligotrophic lakes, while bacterial densities are considerably higher in the former type of lakes. Densities from humic lakes, where the AODC technique has been used, have not been reported previously in the literature. However, the relation between the different lake types agrees with results from extensive reviews of a large number of Polish and Russian lakes (Godlewska-Lipova 1976; Saunders 1980) where the so-called dystrophic lakes occupy an intermediate position between oligo- and eutrophic lakes.

The dense colonization of bacteria on particles found in the present investigation indicates that losses of cells from particulate matter, caused by analytical management, e.g. filtration procedure, should be insignificant. The low share of attached cells in the counts, due to a low frequency of detritus aggregates and not caused by a sparse colonization on these particles, was also confirmed by Hobbie & Wright (1979) while counting bacteria from a wide variety of waters. A predominance of free-living cells, usually more than 90%, has further been reported by Jordan & Likens (1980), Kirchman & Mitchell (1982), Porter & Feig (1980) and Salonen (1977).

Cell shape and size

There seems to exist a connection between cell form and availability of organic matter. In oligotrophic waters a dominance in abundance of coccoid cells is the most common situation (Ferguson & Rublee 1976; Godlewska-Lipova 1976; Wiebe & Pomeroy 1972; Wright 1978), while rod-shaped cells predominate in humic and eutrophic lakes (Godlewska-Lipova 1976). Also within the same water a shift in bacterial shape has been observed due to changed substrate concentrations. In two oligotrophic mountain lakes in northern Sweden, Persson (1977) recorded low densities of bacteria, which were dominated by cocci. After fertilization in one of these lakes there was a strong increase in bacterial density, accompanied by a dominance of rods (Johansson 1980). During enrichment with glucose in the laboratory of marine water containing natural bacteria, Wright (1978) observed a shift from coccoid to larger rod-shaped cells. While culturing a marine *Vibrio* species, Novitsky & Morita (1976, 1977) observed a great reduction during starvation in both size (from larger rods to small cocci) and respiration rate.

The reason for a connection between cell shape and content of organic matter is incompletely known. The cocci, usually small, have a more favourable surface to volume ratio compared to the larger rod-shaped cells, and are hence better adapted to a nutrient-poor environment. The possibility that these small cocci are resting cells with greatly reduced respiration, as was the case in the starvation experiments of Novitsky & Morita referred to above, must also be taken into account. It is possible that the higher share of small cells at the 6 m depth, compared to the surface and bottom waters of Lakes Botjärn and Vitalampa (Table 5), reflect a higher share of resting cells, pehaps caused by the lower availability of substrate.

The seasonal changes in mean volume observed in these lakes probably reflect changes in the growth state of the bacterial populations: bacteria increase their length before they divide (Donachie *et al.* 1976) and a large mean cell volume could thus indicate that a large share of the cells were in division. This was confirmed by Larsson & Hagström (1982) who established a correlation between mean cell volume and the frequency of dividing cells of mixed bacteria isolated from the Baltic and cultured in chemostate at constant temperature and at different growth rates. In Lakes Botjärn and Vitalampa the sharp peaks in mean volumes developing from early May in 1976 and 1977 (Fig. 3) indicate that a large share of the bacterial populations were composed of growing cells. This assumption is also supported by the actual growth of biomass observed during these periods.

Total volume – regulating factors

The development of the bacterial biomass (Fig. 4) in lake Botjärn and Lake Vitalampa during 1975–1977 showed a rather regular and seasonally dependent pattern with the occurrence of one early-summer and one autumn peak. Also in 1974, when counts were made on membrane filters, a large biomass was recorded in October in both lakes. The repeated occurrence of these peaks indicates that also the factors regulating the biomass were rather bound seasonally. Temperature and availability of substrate influence the growth rate. Also the grazing pressure and any contributions of cells from the bottom sediments and from the lake surroundings, are factors potentially important to the biomass development.

When bacterial biomass and mean volume increased at the beginning of the summer, it was interpreted as a response to improved substrate conditions, perhaps due to the relatively large phytoplankton production (Fig. 8). The autochthonously produced substances comprise a substrate source with high affinity as regards bacterial assimilation (Coveney 1982; Herbst & Overbeck 1978; Iturriaga 1981). It is also possible that some of these compounds are involved in the co-metabolism of humic substances (De Haan 1975). Experiments of De Haan (1972, 1975, 1977) and Stabel *et al.* (1979) showed that bacteria in natural waters, in the presence of other assimilable compounds, were able to metabolize substances otherwise not available to them.

Also the increasing temperature and a low grazing pressure during that period should have favoured the growth of bacterial biomass. The elimination of bacteria may partly be caused by zooplankton grazing (Fenchel 1980; Geller & Müller 1981; Nauwerck *et al.* 1980; Peterson *et al.* 1978). Calculations based on filtering rates obtained from *in situ* feeding experiments in Lakes Botjärn and Vitalampa (J.-Å. Johansson, in prep.) and the biomasses of zooplankton (H. Olofsson, in prep.) indicate that the potential grazing pressure was low in May but high in June.

The build-up of the second peak of bacterial biomass during the autumn in all years, coincided with the period of senescence and death of the macrophytes. From August onwards the death of macrophytes could have released large amounts of organic matter (cf. Wetzel 1975). According to F. Eriksson (in prep., pers. comm.) the cumulative elimination of macrophytes, excluding roots, amounted to approximately 320 and 100 kg carbon during August–October in Lake Botjärn and Lake Vitalampa, respectively (mean 1972–1974). Calculated as average values for the total water volumes the elimination rate of macrophytes during September was about 50 mg C · m^{-2} · d^{-1} in the lakes, thus providing a potentially important substrate resource for the bacteria. Also during this period the grazing pressure from zooplankton was calculated to be small.

Compared to the epilimnion, the divergent development of bacterial biomass in the hypolimnion of Lake Botjärn and Lake Vitalampa can be explained by the large differences in environmental factors as regards temperature, oxygen concentration, grazing, availability of substrate, etc. The sedimentation of organic matter into the hypolimnion was probably important, as was evident from the development of the bacterial biomass in Lake Vitalampa. In this lake a large phytoplankton population developed from March or early April in 1976 (Fig. 8). A significant share of this biomass probably sedimented, as both temperature and grazing from zooplankton was low during the spring, implying that mineralization rate was low. Half of the phytoplankton biomass was composed of large *Mallomonas* cells, which ought to have a high sinking rate. The increase in particulate carbon during spring and early summer (Fig. 9), therefore, reflects a supply of autochthonous substrate, causing the rapid growth of bacterial biomass in the hypolimnion from early May (Fig. 4). A corresponding increase of bacteria did not occur in 1975, in spite of a similar large phytoplankton peak. However, this phytoplankton population, which had approximately the same species composition as the population of the following year, developed from early May, when both temperature and the zooplankton population rapidly increased. This implied that decomposition from bacteria and grazing from zooplankton greatly reduced sedimentation of phytoplankton into the hypolimnion.

The high concentrations of particulate carbon in the hypolimnion, developing from August 1975 in Lake Botjärn and from July the following year in both lakes (Fig. 9) cannot be explained from sedimentation of phytoplankton. Neither can the allochthonous inflow of organic matter play a significant role during these periods due to the low or zero runoff in the drainage areas (H. Grip, pers. comm.). The most probable explanation is that a precipitation of iron- and/or aluminium-humic aggregates occurred, caused by the contemporary large increases of the pH values (from about 5.2 to almost 6). These humic precipitates probably are of low nutritional value of the bacteria as compared to the autochthonously produced substances.

Bacterial production

Although the share of external CO_2 in the bacterial production may vary, the seasonal pattern of the dark CO_2 uptake in the epilimnion (Fig. 5) to some extent should reflect changes in the bacterial production rate. The favourable growth conditions during May in all years (see above), correspond to rather high dark CO_2 uptake rates. The large influence of temperature on the dark uptake implied that temperature to a large extent caused high production rates of bacteria in the epilimnion during the summer followed by a rapid decrease during autumn.

The favourable growth conditions created in the hypolimnion of lake Vitalampa in early summer, due to a large supply of organic matter, also corresponded to a rapid production of DIC (Fig. 7) indicating a high bacterial production rate. During this period the chemosynthetic production rate became increasingly important, being very high in this lake at the end of June (Fig. 6).

The production figures in Table 7, all ranging within 15–60 μg C · l^{-1} · d^{-1}, must be regarded as very rough estimates of the total bacterial production. The chemosynthetic production was obviously very high in 1976 in the hypolimnion of the lakes and comprised 20% in Lake Botjärn and 40% in Lake Vitalampa of the total hypolimnic production. The cumulated biomass increase was, on average, 0.5–1 μg C · l^{-1} · d^{-1}. A balance calculation was made for Lake Botjärn as regards the input and output of permanganate-carbon during 1970–1976, resulting in a debt of 44% in the output compared to the quantity transported into the lake (H. Grip, pers. comm.). This corresponded to an elimination of organic carbon of approximately 37 μg C · l^{-1} · d^{-1}. If this deficiency was due to bacterial respiration and if a growth efficiency of 0.4 is assumed, this will imply a bacterial production of 25 μg C · l^{-1} · d^{-1} which is of the same magnitude as the figures in Table 7. However, some of the organic carbon loss within the lake could have been caused by sedimentation and adsorption to the bottom sediment. Also a chemical and photochemical oxidation of the organic matter is possible (Wetzel 1975).

According to Saunders (1980), the average daily

bacterial production in a number of water bodies in the USSR correlated highly with trophic type, ranging from 0.4–15 $\mu g\ C \cdot l^{-1} \cdot d^{-1}$ in oligotrophic and 11–39 $\mu g\ C \cdot l^{-1} \cdot d^{-1}$ in mesotrophic to 36–132 μg $C \cdot l^{-1} \cdot d^{-1}$ in eutrophic waters. In the oligotrophic, clear Mirror lake the yearly bacterial production was, on average, between 1.4 and 3.8 $\mu g\ C \cdot l^{-1} \cdot d^{-1}$ (Jordan & Likens 1980), in the humic Lake Pääjärvi 5–20 $\mu g\ C \cdot l^{-1} \cdot d^{-1}$ in the surface waters and about 1 $\mu g\ C \cdot l^{-1} \cdot d^{-1}$ in the hypolimnion during the summer (Salonen 1981). The bacterial production in the eutrophic Lake Plussee was, on average, 76 $\mu g\ C \cdot l^{-1} \cdot d^{-1}$ (Overbeck 1979). The range of production calculated in Lake Botjärn and Lake Vitalampa may be compared to the values of the mesotrophic waters reported above.

In waters where the autochthonous production is large compared to the input of allochthonous matter, bacteria have the possibility to cover most of their energy requirement from autochthonously produced substances. Thus the ratio between bacterial production and phytoplankton production was usually below 0.4 in different areas of the Baltic (Hagström *et al.* 1979; Rheinheimer 1981) and in lakes with a low content of humic substances though with varying trophic degrees (Jordan & Likens 1980; Riemann *et al.* 1982; Winberg 1980). A ratio of 0.7 or larger has been reported from the humic lakes Pääajärvi (Salonen 1981) and Krugloye (Winberg 1980), as well as from the eutrophic Lake Plussee (Overbeck 1979) which has a considerable input of allochthonous matter in the form of leaves (Albrecht *et al.* 1977). In Lake Botjärn and Lake Vitalampa the phytoplankton production amounted to 4–5 $g\ C \cdot m^{-2}$ both in 1975 and 1976. The elimination of macrophytes during August–October should have supplied between 3–4 $g\ C \cdot m^{-2}$ (F. Eriksson, in prep. and pers. comm.). As the bacterial production was calculated to correspond to more than 7 $g\ C \cdot m^{-2}$ during the growing season, a growth efficiency of 0.4 implies that more than half of the energy requirement of the bacteria must have been taken from allochthonous sources. However, this does not exclude the possibility that autochthonous substances temporarily can play an important role.

Acknowledgements

I thank Dr Mats Jansson and Prof. Birger Pejler for valuable criticism of the manuscript.

References

Albrecht, D., Münster, U. & Stabel, H.-H., 1977. Untersuchungen zum Stoffhaushalt des Plussees. Reprint from: Verh. Ges. Ökol. 7, Jahresversammlung, Kiel 1977: 109–118.

Andersson, I., in press. Bacterioplankton in acidified Lake Gårdsjön. Dev. Hydrobiol.

Blomqvist, P., Heyman, U. & Grundström, R., 1981. The structure of the pelagic ecosystem in Lake Siggeforasjön. Scripta Limnol. Up. A 522: 1–79.

Coveney, M. F., 1982. Bacterial uptake of photosynthetic carbon from freshwater. Oikos 38: 8–20.

Coveney, M. F., Cronberg, G., Evell, M., Larsson, K. & Olofsson, L., 1977. Phytoplankton, zooplankton and bacteria – standing crop and production relationships in a eutrophic lake. Oikos 29: 5–21.

De Haan, H., 1972. Some structural and ecological studies on soluble humic compounds from Tjeukemeer. Verh. int. Ver. Limnol. 18: 685–695.

De Haan, H., 1975. The biological transformation of soluble humic substances in Tjeukemeer, the Netherlands: a preliminary report. In: Pololedo, D. & Golterman, H. L. (eds.). Humic Substances, their Structure and Function in the Biosphere, pp. 63–69. Proc. int. Meet. Humic substances, Nieuwersluis 1972, Pudoc, Wageningen.

De Haan, H., 1977. Effect of benzoate on microbial decomposition of fulvic acids in Tjeukemeer (the Netherlands). Limnol. Oceanogr. 22: 38–44.

Donachie, W. D., Begg, K. J. & Vicente, M., 1976. Cell length, cell growth and cell division. Nature 264: 328–333.

Eriksson, F., Johansson, J.-Å., Mossberg, P., Nyberg, P., Olofsson, H. & Ramberg, L., 1974. Ekosystemets struktur i sjön Vitalampa. Scripta Limnol. Ups. 370: 1–126 (in Swedish, with English summary).

Fenchel, T., 1980. Relation between particle size selection and clearance in suspension-feeding ciliates. Limnol. Oceanogr. 25: 733–738.

Ferguson, R. L. & Rublee, P., 1976. Contribution of bacteria to standing crop of coastal plankton. Limnol. Oceanogr. 21: 141–144.

Geller, W. & Müller, H., 1981. The filtration apparatus of Cladocera: Filter mesh-sizes and their implications on food selectivity. Oecologia 49: 316–321.

Godlewska-Lipowa, W. A., 1976. Bacteria as indicators of the degree of eutrophication and degradation of lakes. Pol. Arch. Hydrobiol. 23: 341–356.

Grip, H., 1982. Water chemistry and runoff in forest streams at Kloten. Division of Hydrology, Department of Physical Geography, Uppsala University, UNGI Report No. 58, 144 pp.

Hagström, A., Larsson, U., Hörstedt, P. & Normark, S., 1979. Frequency of dividing cells (FDC) - a new approach to the determination of bacterial growth rates in aquatic environments. Appl. Envir. Microbiol. 37: 805-812.

Herbst, V. & Overbeck, J., 1978. Metabolic coupling between the alga Oscillatoria redekei and accompanying bacteria. Naturwissenschaften 65: 598.

Hobbie, J. E. & Wright, R. T., 1979. An assessment of quantitative measurements of aquatic microbes. Arch. Hydrobiol. Beih. 13: 85-95.

Iturriaga, R., 1981. Phytoplankton photoassimilated extracellular products; heterotrophic utilization in marine environment. Kieler Meeresforsch. 5: 318-324.

Ivleva, I. V., 1980. The dependence of crustacean respiration rate on body mass and habitat temperature. Int. Revue ges. Hydrobiol. 65: 1-47.

Johansson, J., 1980. Bakterieplankton i Gunillajaure 1978 och 1979. In Kuokkel project progress report 7, pp. 51-57. Inst. of Limnology, Uppsala (in Swedish).

Jordan, M. J. & Likens, G. E., 1980. Measurement of planktonic bacterial production in an oligotrophic lake. Limnol. Oceanogr. 25: 719-732.

Kirchman, D. & Mitchell, R., 1982. Contribution of particle-bound bacteria to total microheterotrophic activity in five ponds and two marshes. Appl. Envir. Microbiol. 43: 200-209.

Krambeck, C., Krambeck, H.-J. & Overbeck, J., 1981. Microcomputer-assisted biomass determination of plankton bacteria on scanning electron micrographs. Appl. Envir. Microbiol. 42: 142-149.

Kuznetsov, S. I. & Romanenko, V. I., 1966. Produktion der Biomasse heterotropher Bacterien und die Geschwindigkeit ihrer Vermehrung im Rybinsk Stausee. Verh. int. Ver. Limnol. 16: 1493-1500.

Larsson, U. & Hagström, Å., 1982. Fractionated phytoplankton primary production, exudate release and bacterial production in a Baltic eutrophication gradient. Mar. Biol. 67: 57-70.

Nauwerck, A., Duncan, A., Hillbricht-Ilkowska, A. & Larsson, P., 1980. In: Le Cren, E. D. & Lowe-McConnell, R. H. (eds.). The Functioning of Freshwater Ecosystems. IBP 22, pp. 251-285. Cambridge Univ. Press. Cambridge.

Novitsky, J. A. & Morita, R. Y., 1976. Morphological characterization of small cells resulting from nutrient starvation of a psychrophilic marine vibrio. Appl. Envir. Microbiol. 32: 617-622.

Novitsky, J. A. & Morita, R. Y., 1977. Survival of a psychrophilic marine vibrio under long-term nutrient starvation. Appl. Envir. Microbiol. 33: 635-641.

Overbeck, J., 1979. Dark CO_2 uptake - biochemical background and its relevance to *in situ* bacterial production. Arch. Hydrobiol. Beih. 12: 38-47.

Persson, G., 1977. Bakterieplankton. In: Kuokkel project progress report 5, pp. 115-116. Inst. of Limnology, Uppsala (in Swedish).

Peterson, B. J., Hobbie, J. E. & Haney, J. F., 1978. Daphnia grazing on natural bacteria. Limnol. Oceanogr. 23: 1039-1044.

Porter, K. G. & Feig, Y. S., 1980. The use of DAPI for identifying and counting aquatic microflora. Limnol. Oceanogr. 25: 943-948.

Ramberg, L., 1976. Relations between phytoplankton and environment in two Swedish forest lakes. Scripta Limnol. Ups. 426: 1-97.

Rheinheimer, G., 1981. Investigations on the role of bacteria in the food web of the Western Baltic. Kieler Meeresforsch. 5: 248-290.

Riemann, B., Søndergaard, M., Schierup, H.-H., Bosselmann, S., Christensen, G., Hansen, J. & Nielsen, B., 1982. Carbon metabolism during a spring Diatom bloom in the eutrophic Lake Mossø. Int. Revue ges. Hydrobiol. 67: 145-185.

Ryhänen, R., 1968. Die Bedeutung der Humussubstanzen im Stoffhaushalt der Gewässer Finnlands. Mitt. int. Ver. Limnol. 14: 168-178.

Salonen, K., 1977. The estimation of bacterioplankton numbers and biomass by phase contrast microscopy. Ann. bot. fenn. 14: 25-28.

Salonen, K., 1981. The ecosystem of the oligotrophic Lake Pääjärvi. 2. Bacterioplankton. Verh. int. Ver. Limnol. 21: 448-453.

Salonen, K., Kononen, K. & Arvola, L., in press. Respiration of plankton in two small, polyhumic lakes. Dev. Hydrobiol.

Saunders, G. W., 1980. Organic matter and decomposers. In: Le Cren, E. D. & Lowe-McConnell, R. H. (eds.). The Functioning of Freshwater Ecosystems, IBP 22, pp. 341-392, Cambridge Univ. Press, Cambridge.

Sederholm, H., Mauranen, A. & Montonen, L., 1973. Some observations on the microbial degradation of humous substances in water. Verh. int. Ver. Limnol. 18: 1301-1305.

Sorokin, J. I., 1965. On the trophic role of chemosynthesis and bacterial biosynthesis in water bodies. Mem. Ist. ital. Idrobiol. 18: 187-205.

Sorokin, Y. I. & Kadota, H., 1972. Microbial production and decomposition in freshwaters. IBP Handbook 23, Oxford, 112 pp.

Stabel, H. H., Moaledj, K. & Overbeck, J., 1979. On the degradation of dissolved organic molecules from Plussee by oligocarbophilic bacteria. Arch. Hydrobiol. Beih. 12: 95-104.

Stainton, M. P., 1973. A syringe gas-stripping procedure for gas-chromatographic determination of dissolved inorganic and organic carbon in fresh water and carbonate in sediments. J. Fish. Res. Bd Can. 30: 1441-1445.

Wetzel, R. G., 1975. Limnology. W. B. Saunders Company, Philadelphia, 741 pp.

Wiebe, W. J. & Pomeroy, L. R., 1972. Microorganisms and their associations with aggregates and detritus in the sea: a microscope study. Mem. Ist. ital. Idrobiol. Suppl. 29: 325-342.

Winberg, G. G., 1980. General characteristics of freshwater ecosystems based on Soviet IBP studies. In: Le Cren, E. D. & Lowe-McConnell, R. H. (eds.). The Functioning of Freshwater Ecosystems, IBP 22, pp. 481-491, Cambridge Univ. Press, Cambridge.

Wright, R. T., 1978. Measurement and significance of specific activity in the heterotrophic bacteria of natural waters. Appl. Envir. Microbiol. 36: 297-305.

Zimmermann, R. & Meyer-Reil, L.-A., 1974. A new method for fluorescence staining of bacterial populations on membrane filters. Kieler Meeresforsch. 30: 24-27.

Relations between production and biomass of phytoplankton in four Swedish lakes of different trophic status and humic content

Ulf Heyman
Institute of Limnology, Box 557, S-751 22 Uppsala, Sweden

Keywords: forest water ecosystems, phytoplankton, biomass, production, growth rate, growth equation, light, nutrients, humic substances

Abstract

Production and biomass values from phytoplankton populations in four different Swedish lakes were analysed. The production in all lakes was directly proportional to biomass during homothermal periods. When the lakes were stratified there was a strong negative relation between specific growth rate and biomass. The data fitted to a logistic density dependent growth equation of the form: $dB/dt = \mu_m B(1\text{-}B \cdot K^{-1})$ where B is the biomass, μ_m the maximum specific growth rate and K the carrying capacity. The equation was used to derive the parameters $\mu \cdot \mu_m^{-1}$ and carrying capacity (the maximum possible biomass). These parameters were then discussed in relation to light climate, phosphorus concentration and humic content.

Introduction

The study of phytoplankton production and biomass has been following two different lines of investigation. One is to study the assimilation versus light curve and from that calculate 'physiological' parameters such as assimilation number and photosynthetic efficiency. These parameters are then compared to environmental parameters and algal species composition (for a review see Harris 1978). This approach is often concerned with time series of measurements in limited areas. The other line of study differences in average values of biomass, chlorophyll and production between different lakes, as in Vollenweider (1976) and Smith (1979). The combination of the two lines of study, i.e. relating time series in different lakes to each other, is seldom used (cf. though Bloesch *et al.* 1977). This state of affairs is probably due to lack of data where production, biomass and environmental factors are measured in sufficient detail and frequency. The biomass is especially hard to measure, since both APT/carbon and chlorophyll/carbon quotients vary too much for their use in studies of population dynamics (cf. Smith 1980).

In order to compare algal populations in different lakes one must find parameters for the populations that can be used in the comparison. Such parameters are, for example, assimilation number, diversity, chlorophyll/carbon quotient, growth rate, nutrient content and nutrient uptake rate. The present paper mainly concerns growth and biomass of algae, wherefore a growth equation describing these parameters is used. The equation is a standard logistic density dependent growth equation and is as such essentially a one species equation and usually not used in productivity studies. The possibility of subdividing the growth into specific growth, maximum specific growth, biomass and carrying capacity (the maximum possible biomass) makes the equation a usable tool when comparisons are made between lakes, since different environmental factors can be directly related to parameters in the equation. Furthermore, equations of this type include elimination of biomass, which otherwise interacts with the effects of nutrients and light.

Hydrobiologia 101, 89–104 (1983).

In the present study, time series of biomass and productivity in four fairly different lakes are analysed. The data are treated both as one whole data set and as separated sets and thus aim at elucidating both the similarities and differences between the different populations. Since the lakes have different phosphorus concentrations and humic content the effect of these two parameters on phytoplankton is of special interest.

Methods

Data from three different investigations from four lakes have been used in the analysis. Measurements were carried out in the years 1975–1976 in Lakes Botjärn and Vitalampa, 1977–1978 in Lake Siggeforesjön and 1976–1978 in Lake Erken. Most measurements were performed in spring, summer and autumn, the measurements in winter being scarce. Most of the data are unpublished but the methods for the measurements of the different parameters are described. For Lakes Botjärn and Vitalampa the methods are the same as those described by Ramberg (1979), for Lake Erken as those described by Boström and Pettersson (1977) and for Lake Siggeforasjön as those described by Blomqvist *et al.* (1981).

Water chemistry was analysed with standard methods, temperature with thermistometers and insolation with various types of light meters. The light attenuation in the water was calculated from Secchi disk transparency (assuming that S.D. = the 1% level) for all lakes but Lake Siggeforasjön, where vertical series of light measurements were used. In L. Botjärn and L. Vitalampa water chemistry was measured only monthly, while in the other lakes it was measured on all sampling occasions.

Primary production in Lake Siggeforasjön was measured with the light and dark bottle C-14 technique and acidification and bubbling after 4 h of incubation. In the other lakes the incubated water was filtered through a 0.2 μm membrane filter after 24 h of incubation. The activity of samples was measured with a liquid scintillation counter. The values from L. Siggeforasjön were transformed to daily values (see Blomqvist *et al.* 1981).

Phytoplankton species composition and volume were analysed by microscopic enumeration under an inverted microscope (Utermöhl 1958). The analysis was performed by the same person (Peter Blomqvist, Inst. of Limnology, Uppsala) in all four lakes, which grants maximal comparability of the biomass estimates. The conversion of biovolumes to phytoplankton carbon was made by multiplying the volumes by 0.11, which is a common value for the relation (cf. Mullin *et al.* 1965, Nalewajko 1966 and Strathmann 1967).

Description of the lakes

Figure 1 gives the geographical position of the lakes. The lakes lie in approximately the same climatic region and the longest distance between any lakes is approximately 350 km. The lakes are quite different both in respect to physical appearance, chemical factors and drainage areas. In Table 1 the physical as well as chemical characteristics of the lakes are given. Lakes Vitalampa and Botjärn are small forest lakes, fairly humic and with their water draining from forests and peat areas. They both lie fairly sheltered from the wind and thus develop shallow epilimnia during summer. The nutrient levels were low. Lake Siggeforasjön is also a forest lake but is influenced by the clay rich soils in its drainage area. Its calcium levels are relatively high, pH is around neutral and alkalinity ~0.1. The

Table 1. Physical and chemical factors in the investigated lakes.

Lake	Area km^2	$\bar{z}$ m	Tot-p[a] $\mu g \cdot l^{-1}$	MRP[a] $\mu g \cdot l^{-1}$	Tot-N[a] $\mu g \cdot l^{-1}$	NO_3-N[a] $\mu g \cdot l^{-1}$	Light factor[a,b]	Colour[a] mg Pt $\cdot l^{-1}$
Botjärn	0.097	3.3	11	5	274	12	0.32	70
Erken	23	9	22	4	550	11	0.22	15
Siggeforasjön	0.76	4.2	16	3	560	150	0.15	60
Vitalampa	0.029	2.8	12	5	220	16	0.37	40

[a] Average for the stratified periods.

[b] Calculated as: $(1-e^{\epsilon z}) \cdot (\epsilon z)^{-1}$, where ϵ is the extinction coefficient and z the mixing depth.

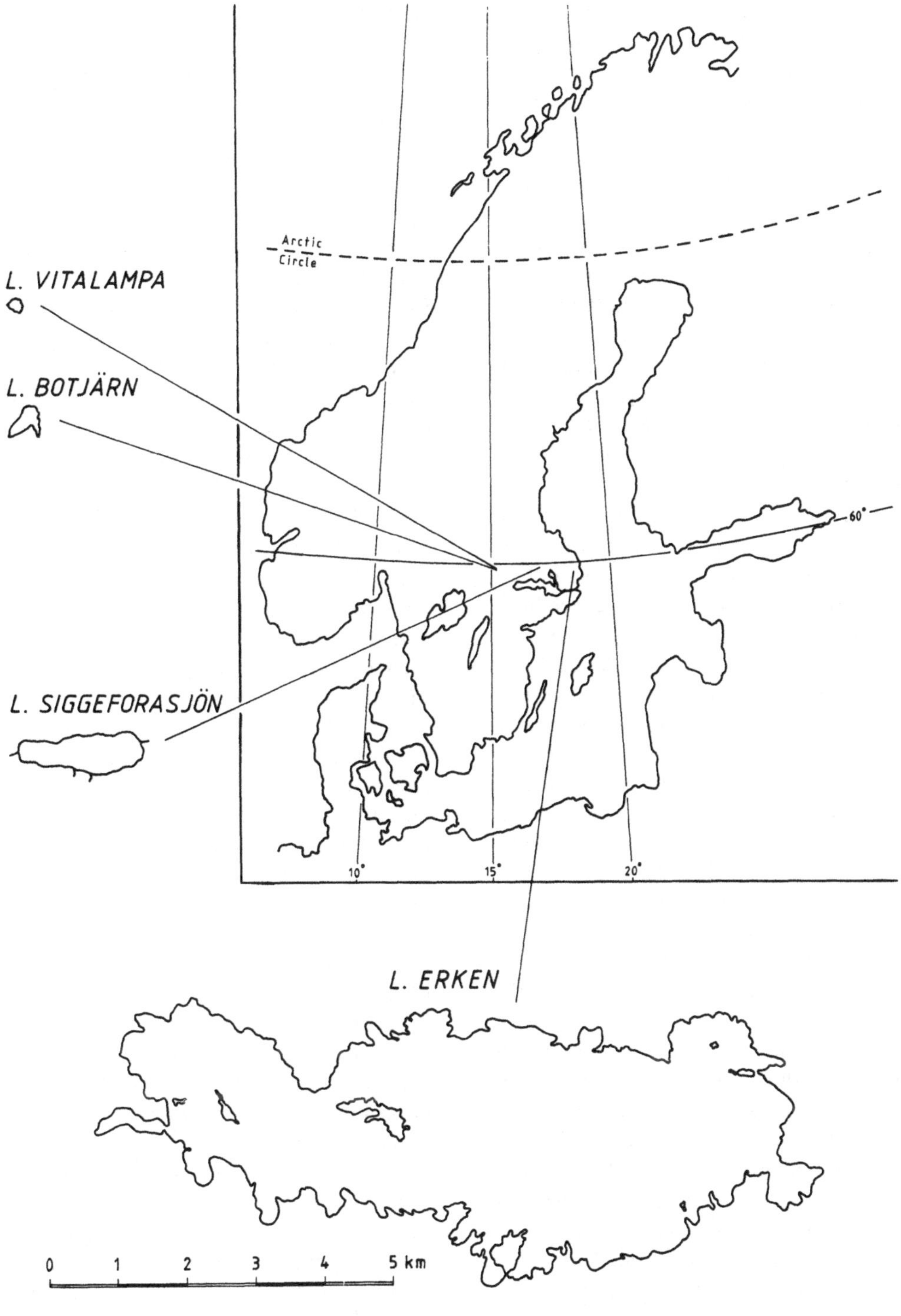

Fig. 1. The positions and sizes of the investigated lakes.

phosphorus and nitrogen levels are higher than in Botjärn and Vitalampa. The lake is medium-sized and since the surroundings are rather flat the depth of the epilimnion was much larger than in L. Botjärn and L. Vitalampa (around 5 m). Lake Erken is a fairly large but shallow lake. It is oligohumic and mesotrophic and the nutrient levels thus the highest of the lakes.

In Table 1 the light climate, i.e. the average light intensity in the epilimnion (Ramberg 1979), is indicated by the factor which should be multiplied by the insolation to obtain the average light. The formula for this factor (Table 1) states that, apart from insolation, the light climate also depends on the attenuation of light and the mixing depth. This means that during periods when the lakes are not stratified the light was highest in the lake with the lowest attenuation coefficient, which was Lake Erken. This, however, only applies when the lakes are free from ice cover and there are many sampling occasions in L. Erken with very low average light intensity. The other lakes were all fairly humic and the attenuation thus higher. Lake Vitalampa, though, had a considerably higher light intensity than L. Botjärn and L. Siggeforasjön. When the lakes stratified the mixing depth became more important and the highest light intensities were found in Lake Vitalampa, the second highest in Lake Botjärn and the lowest in Lake Siggeforasjön (Table 1). It is interesting to note that the light climate in Lake Erken was less than in Lakes Vitalampa and Botjärn due to the much deeper mixing depth (7 m) and thus that the plankton in two of the humic lakes had better light conditions than the plankton in the oligohumic.

The phytoplankton in Lakes Vitalampa and Botjärn consisted of small flagellates from the groups *Chrysophyceae, Cryptophyceae* and *Dinophyceae.* During summer *Chlorococcales* were also important (Ramberg 1979). In Lake Erken diatoms dominated during spring and autumn. In summer *Cryptophyceae, Dinophyceae* and large *Cyanophyta* dominated (Blomqvist, pers. com.; Nauwerck 1963). The species composition of the phytoplankton population in Lake Siggeforasjön was somewhere in between the other lakes, tending towards the forest lakes but with a large contribution of diatoms (Blomqvist *et al.* 1981).

Results

In Table 2 the means, coefficients of variation (cv) and number of data points are given for phytoplankton biomass (as mg carbon · m^{-3}), primary production per square metre of surface area, average primary production per cubic metre of epilimnion water and primary productivity per biomass (P/B). The difference in biomass between the stratified and homothermal periods was small in L. Erken and Siggeforasjön, somewhat larger in L. Botjärn and large in L. Vitalampa. This reflects the more stable stratification in the smaller lakes, which has the result that the spring and autumn biomass peaks largely occurred when the lakes were stratified. Also, in L. Erken, high biomass values are commonly recorded below the ice after the melting of the snowcover in spring (Boström & Pettersson 1977). The carbon assimilation per surface area showed the same differences between the lakes as did biomass. The difference between homothermal and stratified periods was pronounced with much higher productivity during the stratified periods. If the carbon assimilation is expressed per epilimnion volume the difference between periods becomes even more pronounced, while the difference between the lakes decreases. The P/B-quotients did not differ greatly between the lakes, as did the other parameters. This parameter was inversely related to lake trophy so that L. Vitalampa and L. Botjärn had the higher P/B-values. Considering the seasonal variation it is, however, clear that this applied only to the stratified periods, while, during periods of homothermy. Lakes Erken, Siggeforasjön and Vitalampa had about the same P/B and L. Botjärn a considerably lower value.

As can be seen from Table 2 the coefficients of variation within each lake were generally fairly high and the cv-values in L. Erken usually higher than in the other lakes. No parameter showed any tendency of being less variable than the others but it is possible to discern a trend towards higher variability of primary productivity during the stratified periods than during the homothermal. There also seems to be an opposite trend regarding the variability of P/B. The variability of biomass showed no trend and the cv-values in L. Erken and L. Siggeforasjön were lower during stratified periods, while they were higher in Lakes Botjärn and Vitalampa.

The differences between the lakes in biomass and

Table 2. Means and coefficients of variation for the biomass, primary production/surface area, primary production/epilimnetic volume and P/B quotient for the lakes. Values are given for homothermal periods, stratified periods and for all sampling occasions.

Lake	Biomass ($mg\ C \cdot m^{-3}$)			PP/surface ($mg\ C_{ass} \cdot m^{-2} \cdot d^{-1}$)			PP/volume ($mg\ C_{ass} \cdot m^{-3} \cdot d^{-1}$)			P/B (d^{-1})		
	$\bar{x}$	CV%	N	$\bar{x}$	CV%	N	$\bar{x}$	CV%	N	$\bar{x}$	CV%	N
Botjärn												
All samples	17.7	53	42	26.7	52	43	14.2	58	43	0.91	85	42
Homothermal	11.4	19	10	7.3	60	10	2.8	60	10	0.24	56	10
Stratified	19.6	50	32	32.5	30	33	17.7	34	33	1.12	69	32
Erken												
All samples	139	107	50	334	76	44	45.4	76	44	0.54	89	38
Homothermal	134	110	30	274	106	24	30.4	106	24	0.32	52	21
Stratified	145	105	20	406	44	20	63.3	44	20	0.81	74	17
Siggeforasjön												
All samples	49.9	32	50	103	49	50	28.3	57	50	0.61	60	50
Homothermal	55.0	33	21	71.0	44	21	17.4	44	21	0.33	37	21
Stratified	46.1	27	29	126	39	29	36.2	45	29	0.81	44	29
Vitalampa												
All samples	29.9	111	42	31.9	61	43	18.3	64	43	1.15	104	42
Homothermal	13.1	78	9	9.6	70	9	4.0	70	9	0.32	61	9
Stratified	33.5	104	33	37.8	46	34	22.0	46	34	1.37	91	33

primary productivity per square metre of surface area were thus large and the difference could, to some extent, be reduced by expressing the productivity per volume of epilimnetic water. The differences in P/B were comparably small indicating that productivity was largely related to biomass. The higher variability of P/B during stratified periods, however, indicates that the variation in productivity during these periods to a greater extent was dependent on P/B.

In Fig. 2a and 2b the primary productivity per cubic metre is plotted against the phytoplankton biomass for the homothermal and stratified periods. A linear relationship is evident in Fig. 2a, while in 2b the relation, if any, is shadowed by other factors. This confirms the results in Table 2 that P/B during homothermal periods was fairly constant and the productivity thus directly proportional to biomass. A linear regression of primary productivity on biomass gives an R^2-value of 0.88 and a slope of 0.26 mg C_{ass} (mg C)$^{-1}$ · d^{-1}. The regression is naturally very dependent on values from Lake Erken, since these are so much higher than the others, but regressions for each lake separately give similar results. The slopes found in each lake were 0.47, 0.26, 0.22 and 0.20 for Lakes Botjärn, Erken, Siggeforasjön and Vitalampa, respectively. The R^2-values were 0.38, 0.88, 0.26 and 0.50. If the lines are forced through the origo the residual variances do not significantly increase but the slopes become even more similar. The slopes found this way were 0.25, 0.27, 0.31 and 0.27 for the respective lakes and 0.28 if all lakes are included in the regression. The regressions include all measurements made during homothermal conditions and, since the winter samples are scarce, the results are valid only for spring and autumn. The data also include quite a few points with very low average light intensity, which could affect the results because of either a severe light limitation or a large part of the biomass being dormant and thus not possible to include in the calculations. In Fig. 2c the part of Fig. 2a with the low values is shown and points with very low light climate (average light <4 $\mu E \cdot m^{-2} \cdot s^{-1}$) are marked. The good agreement between the lakes is here clearly seen as well as the fact that several of the points which fall somewhat outside the general relationship had low light conditions. The fit to the regressions would thus be better if these points had been excluded. A number of points representing winter data from Lake Erken also fall outside the general trend but since no light values are available, it is impossible to tell whether this is an effect of low light.

It seems clear that the P/B-values during homothermal conditions were remarkably similar between the lakes, remaining constant around 0.3 d^{-1}. This result indicates that the P/B-quotient was independent of biomass but, if a quadratic biomass term is included into the regressions the R^2-values do increase by 0.01 to 0.15 in the lakes. All the quadratic terms were negative but none significant ($p > 0.1$). If, however, all lakes are included in one regression the quadratic term becomes significant ($p < 0.05$) although the R^2 increased by only 0.05.

Temperature is likely to influence the rate of production and, when added to the regression, increased the R^2-values substantially in all lakes except in L. Erken. These regressions are presented in Table 3 where it can be seen that between 67 and 88% of the variation in primary productivity during homothermal conditions could be explained by biomass and temperature. The coefficients for the biomasses changed considerably following the inclusion of temperature, which indicates an interdependence between the two factors. In L. Botjärn and L. Vitalampa the inclusion of temperature led to the coefficient for biomass becoming insignificant. If all lakes are considered, temperature did not have any influence on the primary production.

As can be expected from Fig. 2, the biomass could explain very little of the variation in primary production during stratified periods. The R^2 values for the regressions ranged from 0.002 to 0.05 for the lakes separately while for the whole data set it was 0.23. An inclusion of a quadratic biomass term did not improve the regressions but if temperature was included (Table 3) a significant part of the variation was explained. However, the regression equations presented are in several ways unlikely to reflect a true dependence of primary productivity on temperature and biomass. The very low intercepts and the very high temperature coefficients as well as the varying biomass coefficients make the regressions

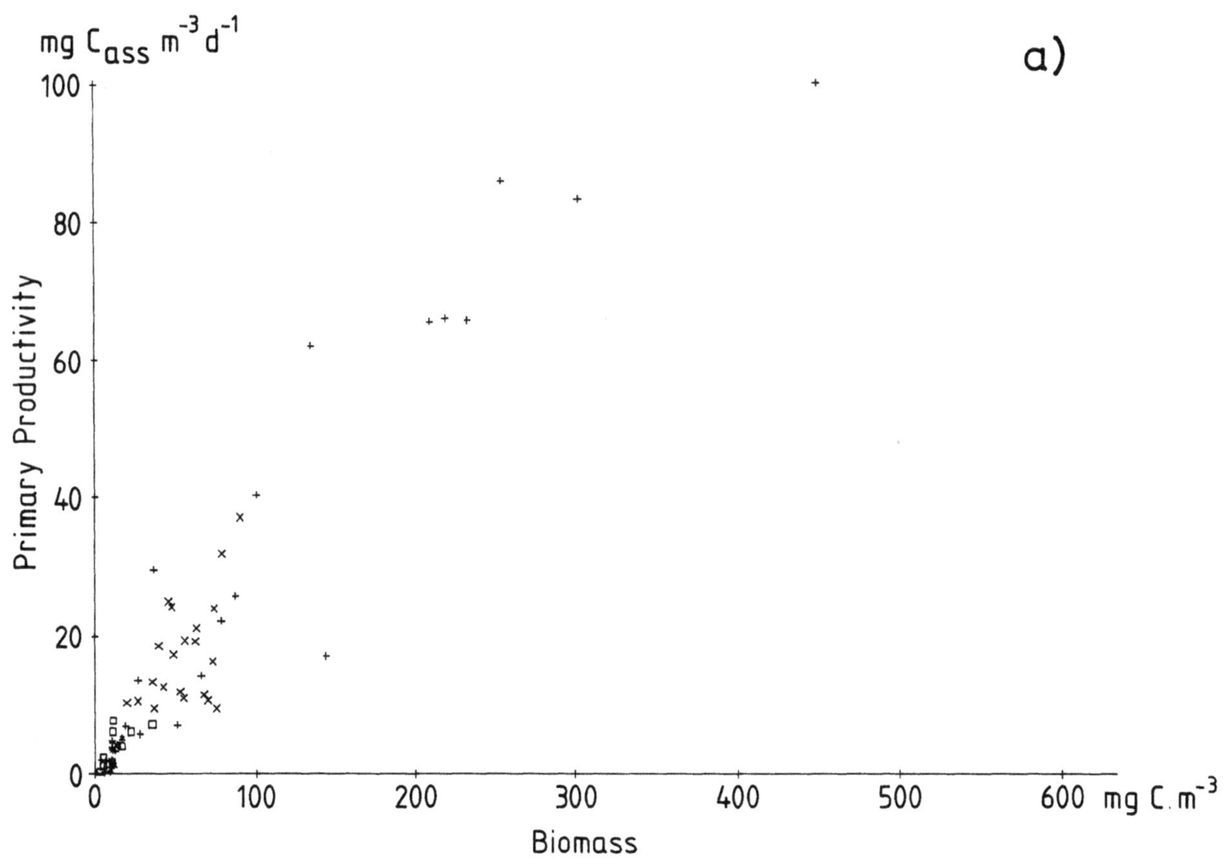

Fig. 2. Relation between primary productivity per volume of epilimnetic water and biomass of phytoplankton. The symbols in the plot denote the different lakes: * denotes L. Botjärn, + L. Erken, × L. Siggeforasjön and □ L. Vitalampa. a) Homothermal conditions.

Table 3. Results of regressions of primary production in temperature and biomass for homothermal and stratified periods in the different lakes and all lakes together.
Equation: PP = a + b_1 biomass + b_2 *temperature + b_3* (Biomass)^2

Lake and period	a	b_1	b_2	b_3*10^3	$R^2\%$	CV%[a]	N
Homothermal							
Botjärn	-3.4	0.20⁻	0.45*		79	31	10
Erken	5.1⁻	0.26**	-0.23⁻		88	34	21
Siggeforasjön	-1.6	0.19*	2.4**		67	27	21
Vitalampa	-0.97⁻	0.10⁻	0.49		80	36	9
All	-0.82⁻	0.36**		-0.27*	91	38	61
Stratified							
Botjärn	no significance				3	35	32
Erken	-22⁻	0.07⁻	5.04⁻		17	45	17
Siggeforasjön	-86*	0.33	6.4**		46	35	29
Vitalampa	-20	0.19*	2.1**		38	38	33
All	-25	0.48**	2.3**	0.75**	39	58	111
All lakes							
All dates	-11*	0.41**	1.49**	-0.49**	56	59	171

- Denotes $p < 0.1$ for the coefficient to be zero.
* Denotes $p < 0.01$ for the coefficient to be zero.
** Denotes $p < 0.001$ for the coefficient to be zero.
[a] Error CV calculated from the residual variance.

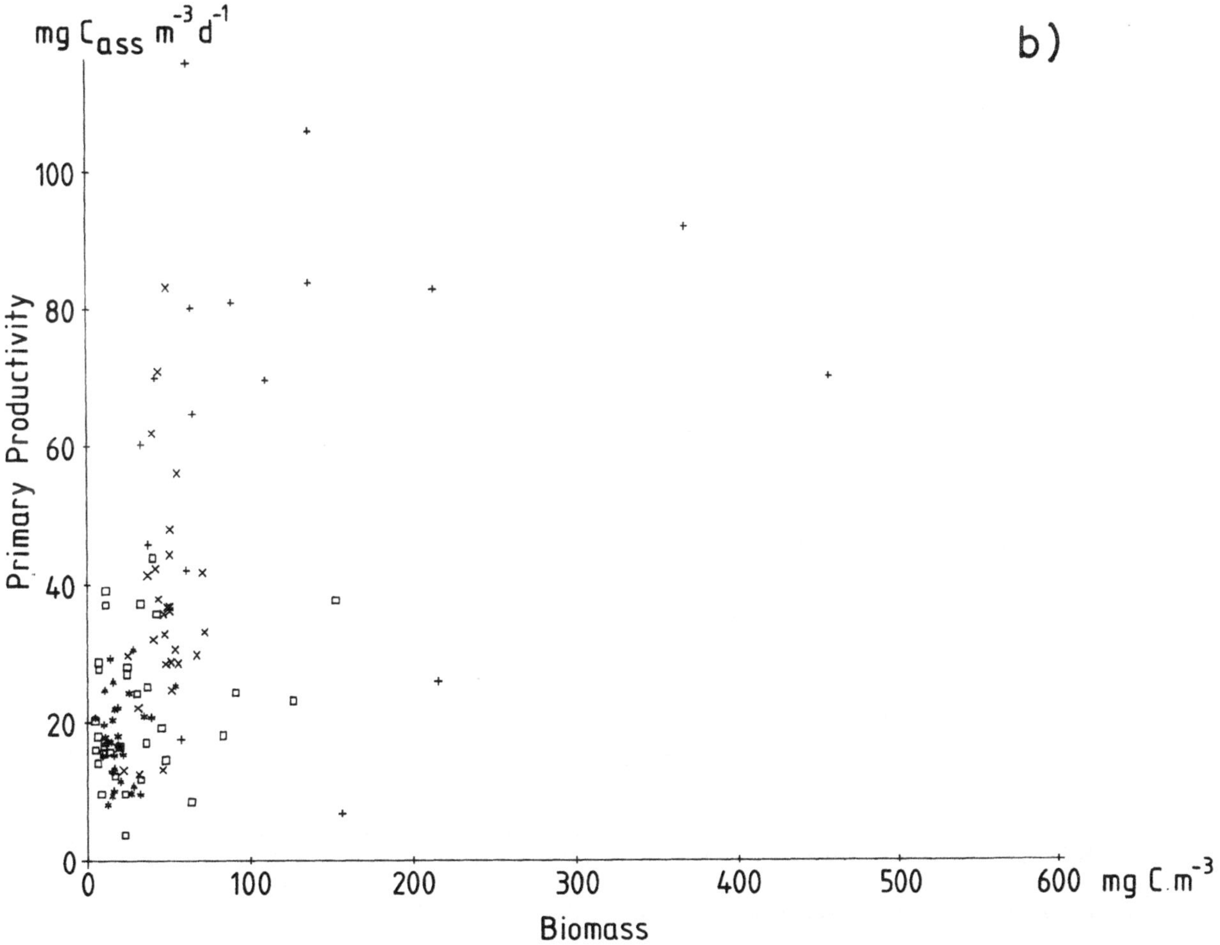

Fig. 2. Relation between primary productivity per volume of epilimnetic water and biomass of phytoplankton. The symbols in the plot denote the different lakes: * denotes L. Botjärn, + L. Erken, × L. Siggeforasjön and □ L. Vitalampa. b) Stratified conditons.

dubious. Probably the primary production during stratified period was to a greater extent governed by variations in the P/B quotient, which, as indicated in Table 2, was more variable during such conditions.

If all values are considered in a regression of primary production on biomass, squared biomass and temperature, 56% of the variation is explained.

Production per biomass could be expected to be temperature dependent and a plot of P/B against mean epilimnetic temperature (Fig. 3) shows that there is a relation. The relation is, however, not very strict and a linear regression only explains about 27% of the variation in P/B. An exponential relation gives, after retransformation, approximately the same residual variance but in view of the general appearance of the plot an exponential relation seems to be preferable. In the plot the different lakes are distinguished with different symbols and it can be seen that several values from L. Vitalampa fall outside the general trend. These values are nearly all from the summer of 1975 but there are no nearby explanations at hand. The main differences between 1975 and 1976 in L. Vitalampa are that the phosphorus concentrations in 1975 were about half of those in 1976 and that there were markedly fewer zooplankton during May and June in 1975 than in 1976. If regressions of P/B on temperature are performed on each lake separately R^2-values range from 0.2 to 0.7 but the resulting exponents vary by a factor 2.4 between the lakes, which makes the regressions found somewhat dubious.

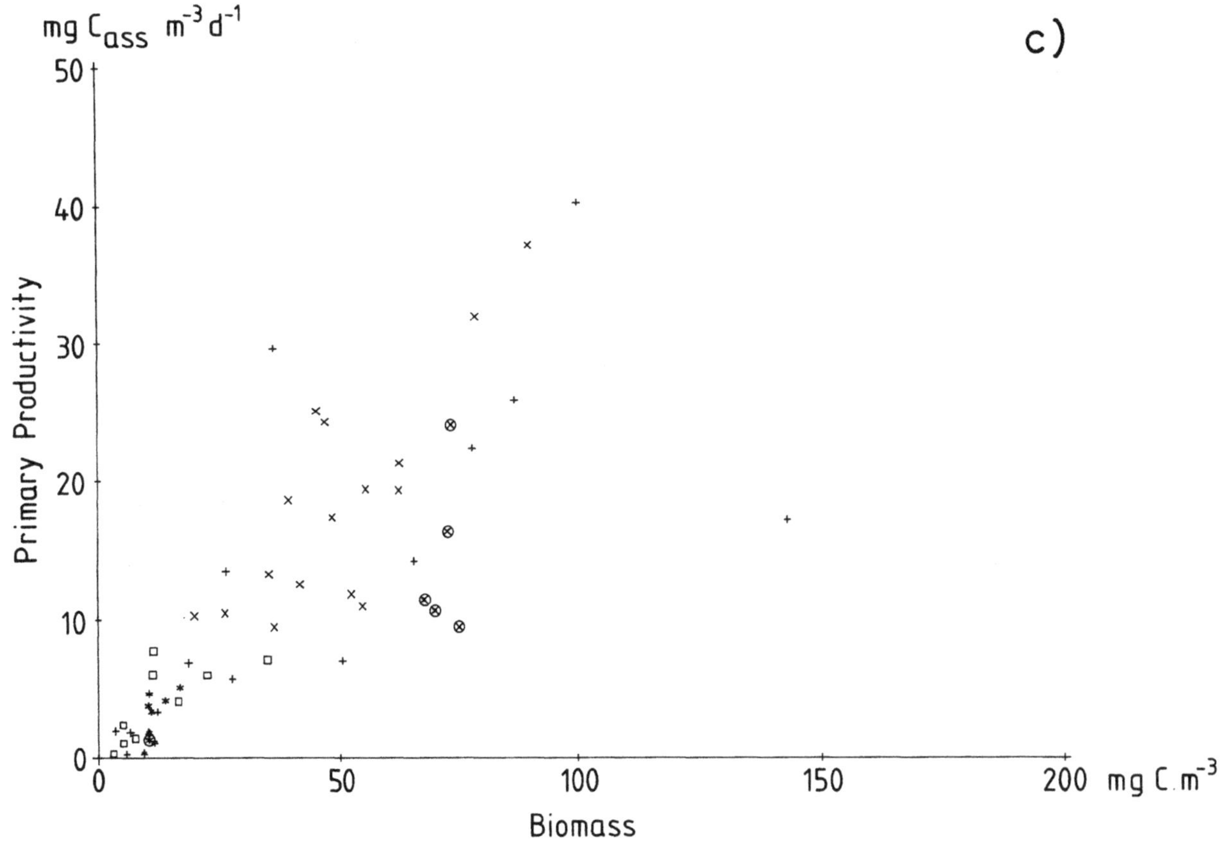

Fig. 2. Relation between primary productivity per volume of epilimnetic water and biomass of phytoplankton. The symbols in the plot denote the different lakes: * denotes L. Botjärn, + L. Erken, × L. Siggeforasjön and □ L. Vitalampa. c) Homothermal conditions but only the lower values.
Circles denote values where the effective light climate was less than 4 $\mu E \cdot m^{-2} \cdot s^{-1}$ as daily average.

The biomass did not significantly affect the P/B quotient during homothermal conditions, except in L. Siggeforasjön, but when the lakes were stratified there was a rather strong negative relation. During these periods regressions of P/B on biomass explained from 26 to 43% of the variation in the individual lakes, while for all lakes it only explained 13%. These results are summarized in Table 4 as multiple regressions of P/B on biomass and temperature. No other parameter improved the regressions to any great extent.

Introducing the logistic density dependent growth equation

Obviously the regressions presented do not explain the variation of primary production during stratified periods. In the homothermal situation, however, the variation of primary production could be explained as a function of biomass and, to some extent, temperature. The unexplained variation corresponded to a coefficient of variation of between 27 and 36%, which, in view of the large possible errors in measurement and biomass estimation, must be regarded as low. The equation for all lakes explained 91% of the variation in volumetric productivity as a function of biomass alone. The P/B quotients, although fairly similar, varied with temperature and to some extent biomass (Table 4). This accounts for the significant negative squared biomass term in a regression where all lakes are considered (Table 3). This type of equation is similar to a logistic density dependent growth equation of the form

Table 4. Regressions of μ on temperature and biomass for homothermal and stratified periods in the different lakes. Equation: $\mu = a + b_1{*}$ temperature $+ b_2{*}$ biomass

Lake and period	a	b_1	$b_2 * 10^3$	$R^2\%$	CV%	N
Homothermal						
Botjärn	-0.04	0.045	- 9.8	72	34	10
Erken	0.34**	0.002	- 0.25	3	54	21
Siggeforasjön	0.11	0.044**	- 3.4**	73	20	21
Vitalampa	0.11	0.048°	-11	60	44	9
Stratified						
Botjärn	3.2**	-0.06	-48**	41	54	32
Erken	-0.09	0.08	- 2.6°	53	54	17
Siggeforasjön	-1.1°	0.14**	- 7.9°	43	35	29
Vitalampa	-0.80	0.15°	-11°	45	70	33

Symbols as in Table 3.

(1) $dB/dt = \mu \cdot B \cdot ((1-B) \cdot K^{-1})$ (cf May 1976)

where μ_m is the maximum specific growth rate and K the carrying capacity of the system. If both sides of the equation are divided by B one obtains

(2) $dB/dt = P \cdot B^{-1} = \mu = \mu_m \cdot (1-B) \cdot K^{-1})$

where μ is the specific growth rate. The P/B quotients given in this study are not directly applicable to the equation, since the exposure times are comparatively long and B thus likely to vary during the uptake experiments. It has been suggested (Peterson 1978) that $\mu = (\Delta t \cdot \ln((B + C_{ass}) \cdot B^{-1}))^{-1}$ could be used but this formula assumes that all the assimilated carbon is used to increase biomass. Contrary to this assumption, there is usually very little biomass increase compared to the carbon assimilation within the lakes. This applies also when the grazing pressure is negligible. Peterson (1978) found that the quotient between the change in particulate carbon and the assimilated ^{14}C ($\Delta PC \cdot {}^{14}C^{-1}$-ass) increased linearly with the net growth of phytoplankton. A net growth rate of 0.25 d^{-1} would, according to Peterson, result in a $\cdot PC \cdot {}^{14}C^{-1}$-ass-ratio of ~0.5. These findings refute Peterson's calculation of μ. In view of the usually very small changes in biomass in the studied lakes μ is in the following calculated as $\mu = P/B$. This might introduce an error but the error will probably be smaller than the error encountered when the equation used by Peterson is employed.

A negative relation between μ and biomass is clearly stated in equation (2) and the negative covariation between specific growth rate and biomass during stratified periods may well be an effect of a relation similar to equation 1. In the equation, μ_m should be strongly temperature-dependent because of its relation to enzymatic processes and this dependence has been discussed by several authors (i.e., Eppley 1972; Goldman & Carpenter 1974). Eppley (1972) constructed an emvelope-type curve from algal culture data that fitted the equation $\mu_m = 0.851 \cdot (1.066) \cdot T$. Since Eppley's equation is based on changes in biomass and the present study deals with carbon assimilation, the equation cannot be used directly because of the carbon loss from respiration and excretion. Assuming that these losses are 20% of the carbon uptake would give the

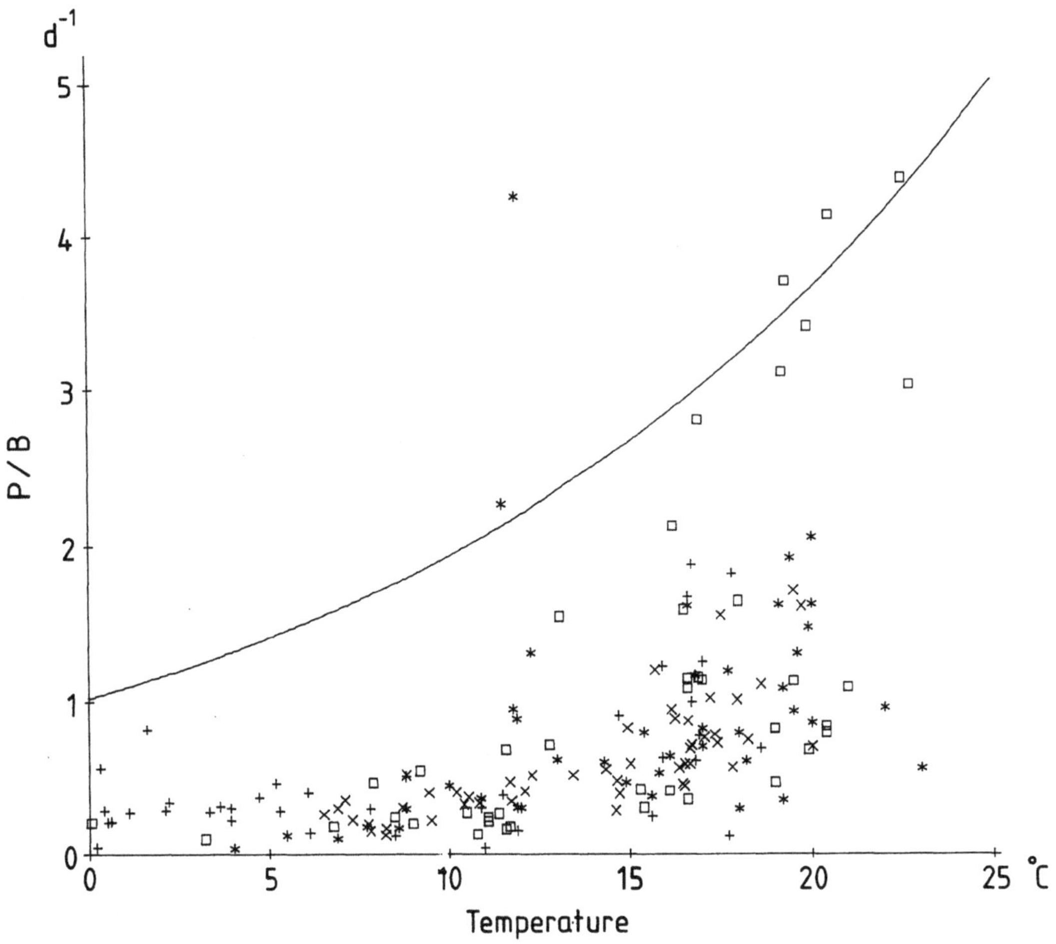

Fig. 3. Relation between production per biomass and temperature. Symbols are the same as in Fig. 2. The line describes the equation $\mu_m = 1.02\ (1.066)$ temperature (see text).

equation $\mu_m = 1.2 \cdot 0.851 \cdot (1.066) \cdot T$, which is used below. In Fig. 3, the line describing the equation is included. The value 1.066 in Eppley's equation yields a Q_{10} value of about 1.88, which is much lower than the Q_{10} of about 2.7 that is implicated by the regression of ln(P/B) on temperature in this study. The material at hand, however, is not good enough to construct a new equation and if regressions of μ on μ_m are compared with the regressions of ln(P/B) on temperature the difference in R^2 is not very large.

Now, according to the equation (2), the quotient between μ and μ_m should be negatively dependent on the biomass divided by the carrying capacity. If the biomass is a constant fraction of K the quotient between μ and μ_m remains constant. This seems, more or less, to be the case within each lake during homothermal conditions but, according to the significant quadric biomasss term when all lakes are considered, different quotients in the different lakes must be expected. When the lakes were stratified the biomass covaried negatively with primary productivity, which indicates that the quotient between biomass and carrying capacity varied.

Regression analysis shows that a fairly large part (10–43%) of the variation in $\mu \cdot \mu_m^{-1}$ is explained by B. Furthermore, the unexplained standard deviations for these regressions were much less than in the multiple regressions of μ on temperature and biomass. Obviously, the temperature dependence is better described by the growth equation.

Equation (2) can be rewritten to

(3) $$1 - (\mu \cdot \mu_m^{-1}) = B \cdot K^{-1}$$

Table 5. Regressions of $(1-\mu \cdot \mu_m^{-1})$ on biomass for the stratified periods in the different lakes.
Equation: $1-(\mu \cdot \mu_m^{-1}) = a + b * B$

Lake	a	b*10³	R²%	CV%
Botjärn	0.53**	7.9*	29	18
Erken	0.58**	1.1*	43	21
Siggeforasjön	0.62**	2.5°	10	13
Vitalampa	0.55**	3.7*	34	27
All	0.65**	1.0**	16	22

Symbols as in Table 3.

and thus, to evaluate the quotient between biomass and carrying capacity, a regression of $1-(\mu \cdot \mu_m^{-1})$ on biomass should be made. Such regressions for the stratified periods are presented in Table 5. Before the regressions were made, 7 points with $\mu \cdot \mu_m^{-3} > 0.9$ were deleted from the data, since such high quotients lead to negative or extremely high K-values. The intercepts of the equations, which theoretically should be zero, are fairly high but do not vary much between the lakes. The high values indicate that μ seldom was higher than $\sim 0.5 \cdot \mu_m$, which also is evident from Fig. 3. Since the carrying capacity, due to different trophic status of the lakes, should be different in the different lakes, it was expected that the slopes of regression lines would vary widely between the lakes. This difference between the lakes also causes the bad fit of an equation including all the lakes. To achieve estimates of the carrying capacity in the different lakes the regressions above were recalculated with the lines forced through the origo. The slopes thus found were 0.029, 0.0035, 0.015 and 0.011 for Lakes Botjärn, Erken, Siggeforasjön and Vitalampa, respectively. The carrying capacity would then be the inverted slopes which for the respective lakes will be 34.5, 286, 67 and 91 mg C $\cdot$ m^{-3}.

In Table 6 the means and coefficients of variation for the measured parameters μ and B as well as for the parameters μ_m, $\mu \cdot \mu_m^{-1}$ and K in equation (2) are given for each lake and period. K was calculated directly from the equation as $K = B \cdot (1-(\mu \cdot \mu_m^{-1}))^{-1}$. The parameter $\mu \cdot \mu_m^{-1}$ can be considered as a measure of the specific growth rate standardized to constant temperature. Thus, the difference in μ, found between stratified and homothermal periods, was much reduced when μ was divided by μ_m. Also the difference in μ found between lakes for stratified periods as well as for the whole seasons, was almost removed. In most cases the cv. is significant-

Table 6. Means and coefficients of variation for the parameters of the growth equation in the different lakes. Values are given for homothermal periods, stratified periods and for sampling occasions.

Lake	μ d^{-1}		μ_m d^{-1}		$\mu \cdot \mu_m^{-1}$		Biomass mg C $\cdot$ m^{-3}		K mg C $\cdot$ m^{-3}		N
	$\bar{x}$	CV%	$\bar{x}$	CV%	$\bar{x}$	CV%	$\bar{x}$	CV%	$\bar{x}$	CV%	
Botjärn											
All samples	0.79	66	2.8	28	0.27	57	18.2	51	25.4	47	40
Homothermal	0.23	56	1.8	17	0.12	48	11.4	19	13.1	22	10
Stratified	0.97	48	3.15	18	0.31	45	20.4	47	29.5	37	30
Erken											
All samples	0.54	89	1.96	41	0.27	63	125	94	162	83	38
Homothermal	0.32	52	1.31	18	0.25	58	117	102	154	97	21
Stratified	0.81	74	2.77	15	0.28	69	135	88	172	69	17
Siggeforasjön											
All samples	0.61	60	2.53	24	0.23	42	49.9	32	64.5	28	50
Homothermal	0.33	37	1.89	14	0.17	31	55.0	33	66.0	31	21
Stratified	0.81	44	10	3.0	0.27	38	46.1	27	63.4	26	29
Vitalampa											
All samples	0.81	93	2.65	30	0.28	75	32.1	104	43.2	87	37
Homothermal	0.32	61	1.68	21	0.19	48	13.1	78	15.9	71	9
Stratified	0.96	83	2.96	22	0.31	74	38.2	94	52.0	75	28
All											
All samples	0.68	79	2.49	32	0.26	61	55.6	129	72.8	116	165
Homothermal	0.31	49	1.64	23	0.19	56	63.1	131	80.4	130	61
Stratified	0.90	63	2.99	17	0.29	57	51.2	126	68.4	103	104

ly reduced by the division by μ_m, again indicating the temperature dependence of μ. Since the $\mu \cdot \mu_m^{-1}$ quotients were fairly similar between the lakes, the relation between biomass and carrying capacity must also be similar and around 0.7. The average K-values differ from those found from the regression slopes, especially in Lakes Erken and Vitalampa. This is due to non-linearities of the relations between $(1-\mu) \cdot \mu_m^{-1}$ and B. The most likely cause of the non-linearity is that the biomass estimates are incorrect. Both the inclusion of non-viable biomass and varying carbon content per cell volume would give the type of non-linearity found, because of the division by biomass in the specific growth estimate. The found non-linearity does to some extent refute the use of equation (2) but use of the equation explains more of the variation in μ and also has advantages from a theoretical viewpoint.

Discussion

The algae in lakes presented in this study evidently function in different ways depending on whether the lakes are stratified or not. During periods of homothermal conditions, when light and temperature are low, the growth rate remains fairly constant while the biomass varies. In periods with stratification the growth rate seems more variable and strongly dependent on biomass. This is a common result in productivity studies (i.e., Rodhe *et al.* 1956; Bloesch *et al.* 1977; Kalff & Knoechel 1978) and also in chemostat experiments (cf. Ahlgren 1980). This result is, however, usually not used in phytoplankton growth equations. Formally, a growth equation as equation (1) cannot be applied to an assemblage of algae, since it is essentially a one species equation. As a rough tool while waiting for more exact relations it might, however, be used as is the Monod type equation. Both the maximum specific growth rate and the carrying capacity are likely to be different for different species. The advantage of using a growth equation is that it provides a possibility to subdivide the phytoplankton response to different environmental parameters. Temperature thus can be considered to act only on μ_m (see however Goldman & Carpenter 1974), the parameters $\mu \cdot \mu_m^{-1}$ and K thus being temperature independent. The carrying capacity is a function of nutrient and energy supply. In lakes where phytoplankton biomass has no significant impact on the light conditions, as in the present study, light cannot be included in the carrying capacity. In such lakes light, if limiting the production, will only affect the rate of increase to a certain biomass and not the level of the maximum possible biomass. This restricts the use of the equation to periods with relatively high light, e.g. during stratified conditions. The parameter $\mu \cdot \mu_m^{-1}$ is either a measure of how large the biomass is compared to K, or a measure of light limitation. Thus, normally, grazing and other elimination factors will act on $\mu \cdot \mu_m^{-1}$.

The most striking result of the present study is how similar production and biomass relate to each other in the different lakes. For periods of homothermal conditions it was possible to fit all data into one single regression line for production on biomass. The relation between P and B differed very little between the lakes although the mean biomass and productivity varied by an order of magnitude. The growth rates thus were fairly similar between the lakes, although L. Botjärn differed from the rest. It is interesting to note that there were considerably greater differences in $\mu \cdot \mu_m^{-1}$ than in μ. This might indicate that a specific growth rate of about 0.3 d^{-1} is needed to sustain the phytoplankton populations regardless of the temperature. The elimination of assimilated carbon (respiration, excretion, sedimentation, mortality and grazing) should thus also be around 0.3 d^{-1} even though the grazing pressure during these periods was normally low. The overall impression of the phytoplankton dynamics during homothermal periods is thus that the populations achieved a high biomass compared to the carrying capacity. To compensate for the elimination and thereby to sustain the biomass a growth rate of 0.2–0.3 d^{-1} seemed to be needed.

In contrast to the situation during homothermal periods specific growth rates during periods of stratification differed more between the lakes. The difference between the lakes becomes large if the high values, especially from L. Vitalampa, are included in the data (cf. Table 2 and Table 6). With the high values deleted, average $\mu \cdot \mu_m^{-1}$ quotients become very similar and phytoplankton normally seems to have a specific growth rate that corresponds to a $\mu \cdot \mu_m^{-1}$ value of ~0.3. The quotients were, however, fairly variable within each lake (Table 6) and the insignificant trend of $\mu \cdot \mu_m^{-1}$ quotients, being lower in Lakes Siggeforasjön and

Erken than in Lakes Botjärn and Vitalampa, becomes very significant if all values are taken into account. The quotients in Lakes Vitalampa and Botjärn would then be 0.41 and 0.39, respectively.

This study covers four lakes which in many respects are different but their phytoplankton seems to behave very similarly in terms of productivity and biomass. Differences seem to be restricted to the levels of carrying capacity and possibly, during stratification, $\mu \cdot \mu_m^{-1}$. Other differences such as the primary production per water volume are only functions of those mentioned. Comparisons of water chemistry and light climate between the lakes can only be made for the stratified periods due to lack of data. The light climate (Table 1) varied considerably between the lakes in the order Lakes Siggeforasjön, Erken, Botjärn and Vitalampa from lowest to highest light factor. This is the same sequence that the $\mu \cdot \mu_m^{-1}$ quotient followed and it may not be unlikely that a high energy input leads to a high carbon turnover rate (cf. Hobson 1974) and thus to a high productivity and elimination. Even though the $\mu \cdot \mu_m^{-1}$ values, because of the high values in L. Vitalampa, are somewhat unreliable, it is clear that there was not any inhibition of the specific growth rate due to humic substances in the water. Coloured humic substances naturally affect the extinction coefficient and thus the light climate but since mixing depth is as important as the extinction for the light climate, the possible relation between $\mu \cdot \mu_m^{-1}$ and light in the studied lakes is not related to humic content.

As noted above, the carrying capacity should be a function of nutrient supply. Since it is a measure of possible biomass it is independent of the turnover rate of nutrients. The most likely nutrient to limit K is phosphorus but considering that the quotient between average N and average P in three of the lakes (cf. Table 1) was below 25, nitrogen may well have been limiting on some occasions. A comparison of the means of K and the means of total phosphorus concentration (Tables 1 and 6) shows that K increases with increasing P. Since several values from L. Vitalampa are deleted in the calculation of mean K, also the corresponding phosphorus values should be deleted in the comparison. This leads to an increase of the average P concentration from 12 mg $\cdot$ m^{-3} to 14 mg $\cdot$ m^{-3} in L. Vitalampa, which makes the P and K values even more comparable. In Lakes Erken and Siggeforasjön it is possible to compare individual values of P and K, since phosphorus in these lakes was measured on all sampling occasions. The R^2 values for the regressions of K on P were 0.27 and 0.14 for the respective lakes, which is fairly low but significant. The values are also higher than comparable values for regressions of biomass and primary productivity on total P. K thus seems to be the parameter most dependent on available nutrients and probably the regressions would have had explained more of the variation in K if P was the only limiting factor and if a measure of available P could be used instead of the concentration of total P.

The quotients between K and P are not the same in the different lakes. If the K-values found from the regressions of $1-(\mu \cdot \mu_m^{-1})$ on biomass (see above), which are probably somewhat better due to the lesser effect of small and possibly light-limited values, are multiplied by a cell P:C ratio an estimate of the amount of available P will result. Using a quotient of 1 : 62 (Ulén 1977) results in 0.6, 4.6, 1.1 and 1.5 mg available P $\cdot$ m^{-3} in Lakes Botjärn, Erken, Siggeforasjön and Vitalampa, respectively. Expressing these values as percentages of total P, yields 5, 21, 7 and 11% for the respective lakes. This series is inversely correlated to the humic content of the lakes, since L. Erken is the least humic, L. Vitalampa fairly humic and Lakes Siggeforasjön and Botjärn more humic than L. Vitalampa. Apparently the amount of humic substances covaries with a decreasing relative amount of available P, which also has been indicated in other studies (Sakamoto 1971; Steeman-Nielssen 1978). Jackson & Heckey (1980), in contrast, indicated that the inhibition of primary productivity in humic reservoirs was due to iron deficiency. This possibility cannot be refuted by the present study although the relationships between phytoplankton parameters and phosphorus indicate that phosphorus is the most important nutrient.

The phytoplankton in the four lakes exhibited very similar production characteristics. The only large difference between the populations, apart from the species composition, was the difference in carrying capacity, which was dependent on the phosphorus concentration and humic content. During homothermal conditions low light climate sometimes limited production and also light may have, to some extent, influenced the $\mu \cdot \mu_m^{-1}$ ratio during stratified conditions. The light climate was, however, not correlated to the humic content but primarily to the mixing depth.

The present paper advances the use of a logistic density dependent growth equation to analyse results from studies of primary productivity and biomass of phytoplankton. By using such an equation it is possible to relate specific production to biomass, which otherwise has to be done by the modelling of nutrient concentrations. Instead of the growth equation, introduction of a nutrient uptake equation is then needed (usually of a Michaelis-Menten type). Such equations have been used for many years (cf. Patten 1968) and usually seem to fit data adequately. The equations are, however, not readily applicable to productivity data since nutrient turnover rates, available nutrient concentrations, half-saturation constants and so forth are not known. A growth equation is less exact but gives parameter values that can be directly compared to normal limnological data. The double quotients as well as the unreliable temperature dependence of μ_m does, however, limit the usability of the equation. Values can easily, as in the present study, fall outside the defined range of the equation. The equation is, however, rather insensitive to changes in the temperature dependence as long as relative comparisons are made. If, for example, the temperature dependence found from the regressions of ln(P/B) on temperature is used instead of Eppley's (1972) equation, giving the relation $\mu_m = 0.851 \cdot (1.112)$ temperature, the relations between the lakes concerning all parameters would be the same. The equation presupposes that the growth rate is regulated by the biomass but there is no indication as to what is regulating the biomass, although the equation is constructed for self-regulation. It is interesting to note that the populations in this study always had biomasses considerably lower than their carrying capacities. There thus seems to be definite advantages for the planktonic system, including the grazers, to keep the phytoplankton biomass fairly low. According to the equation the low biomass will increase the primary productivity of the system, since the second derivative of the equation is zero when $B \cdot K^{-1} = 0.5$, and evolution might have favoured such a development (cf. also Ramberg 1980).

Acknowledgement

I wish to express my gratitude to Peter Blomqvist, Bengt Boström, Jan-Åke Johansson, Hans Olofsson and Kurt Pettersson for letting me use their data. My thanks also to Dr. Arnold Nauwerck for valuable comments on the manuscript.

References

Ahlgren, G., 1978. Growth of Oscillatoria agardhii in chemostat culture. 2. Dependence of growth constants on temperature. Mitt. int. Ver. Limnol. 21: 88-102.

Bloesch, J., Stadelman, P. & Buhrer, H., 1977. Primary production, mineralization and sedimentation in the euphotic zone of two Swiss lakes. Limnol. Oceanogr. 19: pp. 511-526.

Blomqvist, P., Heyman, U. & Grundström, R., 1981. The structure of the pelagic ecosystem in Lake Siggeforasjön. Scripta Limnol. Ups. A: 522.

Boström, B. & Petersson, K., 1977. The spring development of phytoplankton in Lake Erken, Freshwat. Biol. 7: 327-335.

Eppley, R. W., 1972. Temperature and phytoplankton growth in the sea. Fish Bull. U.S.A. 70: 1063-1085.

Goldman, J. C. & Carpenter, E. J., 1974. A kinetic approach to the effect of temperature on algal growth. Limnol. Oceanogr. 19: 756-766.

Harris, G. P., 1978. Photosynthesis and growth: The physiological ecology of phytoplankton. Arch. Hydrobiol. Beih. 10: 1-171.

Hobson, L. A., 1974. Effects of interactions of irradiance, daylength and temperature on division rates of three species of marine unicellular algae. J. Fish. Res. Bd. Can. 31: 391-395.

Jackson, T. A. & Heckey, R. E., 1980. Depression of primary productivity by humic matter in lake and reservoir waters of the boreal forest zone. Can. J. Fish. aquat. Sci. 37: 2300-2317.

Kalff, J. & Knoechel, R., 1978. Phytoplankton and their dynamics in oligotrophic and eutrophic lakes. Annu. Rev. Ecol. Syst. 9: 475-495.

May, R. M., 1976. Models for single populations. In: May, R. M. (ed.). Theoretical Ecology. Blackwell Sci. Publ.: 4-25.

Mullins, M. M., Sloan, P. R. & Eppley, R. W., 1965. Relationship between carbon content, cell volume and area in phytoplankton. Limnol. Oceanogr. 11: 307-311.

Nalewajko, C., 1966. Dry weight, ash and volume data for some freshwater planktonic algae. J. Fish. Res. Bd Can. 23: 1285-1288.

Nauwerck, A., 1963. Die Beziehungen zwischen Zooplankton und Phytoplankton im See Erken. Symb. Bot. Ups. 17: 163 pp.

Patten, B. C., 1971. Systems analysis and simulation in ecology. Acad. Press, London.

Peterson, B. J., 1978. Radiocarbon uptake: its relation to net particulate carbon production. Limnol. Oceanogr. 23: 179-184.

Ramberg, L., 1979. Relations between phytoplankton and light climate in two Swedish forest lakes. Int. Revue. ges. Hydrobiol. 64: 749–782.

Ramberg, L., 1980. A population dynamics model of Oocystis parva (Chlorophyceae). Arch. Hydrobiol. 89: 119–134.

Rodhe, W., Vollenweider, R. A. & Nauwerck, A., 1958. The primary productivity and standing crop of phytoplankton. In: Buzzati-Traverso, A. A. (Ed.) Perspectives in marine biology. A symposium held at Scripps Inst. Oceanogr. Univ. Calif., March 24–April 2, 1956: 299–322.

Sakamoto, M., 1971. Chemical factors involved in the control of phytoplankton production in the Experimental Lakes Area, Northwestern Ontario. J. Fish. Res. Bd Can. 28: 203–213.

Smith, R. A., 1980. The theoretical basis for estimating phytoplankton production and specific growth rate from chlorophyll, light and temperature data. Ecol. Modelling 10: 243–264.

Smith, V. H., 1979. Nutrient dependence of primary productivity in lakes. Limnol. Oceanogr. 24: 1051–1064.

Steeman-Nielssen, E., 1974. Growth of unicellular alga Selenastrum capricornutum as a function of P. with some information also on N. Verh. int. Ver. Limnol. 20: 38–42.

Strathmann, R. R., 1969. Estimating the organic carbon content of phytoplankton from cell volume and plasma volume. Limnol. Oceanogr. 12: 411–418.

Ulén, B., 1977. Seston and sediment in L. Norrviken. I Seston composition and sedimentation. Scripta Limnol. Ups. 446.

Utermöhl, H., 1958. Vervollkommnung der quantitativen Phytoplanktonmetodik. IAL Mitt. 9.

Vollenweider, R. A., 1976. Advances in difining critical loading levels for phosphorus in lake eutrophication. Mem. Ist. ital. Idrobiol. 33: 53–83.

Primary production and phytoplankton in two small, polyhumic forest lakes in southern Finland

L. Arvola
Lammi Biological Station, University of Helsinki, SF-16900 Lammi, Finland

Keywords: forest water ecosystems, phytoplankton, primary production, polyhumic lake, *Chlamydomonas, Mallomonas*

Abstract

Primary production and phytoplankton in polyhumic lakes showed a very distinct seasonal succession. A vigorous spring maximum produced by *Chlamydomonas* green algae at the beginning of the growing season and two summer maxima composed mainly of *Mallomonas caudata* Iwanoff were typical. The annual primary production was ca. 6 g org. C · m^{-2} in both lakes. The mean epilimnetic biomass was 1.1 in the first lake and 2.2 g · m^{-2} (ww) in the second one. The maximum phytoplankton biomass, 14 g · m^{-2}, was observed during the vernal peak in May.

Introduction

Small, brown-water forest lakes form an ecological environment with several extraordinary features. The water is often acidic; the concentration of soluble and colloidal humic compounds is usually high; transmission of light is low; turbulent water currents are weak and anoxia in the hypolimnion is common. The ecology of phytoplankton in these small lakes is as little understood as these ecosystems generally, although there are thousands of such lakes in northern Europe. Only some notes of such small lakes and their phytoplankton have been published in Finland (Meriläinen 1970; Lehmusluoto & Ryhänen 1972; Ilmavirta 1980).

In this study the seasonal succession of primary production and phytoplankton in two small, polyhumic forest lakes in southern Finland during one year (1979) is described, and the factors which seem to affect the succession are discussed.

Study area

The study area is situated in Lammi commune in southern Finland (Fig. 1). The small lakes studied are Lake Nimetön and Lake Horkkajärvi. Some morphometric and physico-chemical properties of these lakes are shown in Table 1.

Both lakes are surrounded by tall forest, which offers an effective protection from wind. Protected location, small surface area, and relatively great depth cause meromixis with an anoxic hypolimnion. In summer the oxygenated layer may be only 1–2 meter thick.

The shoreline of Nimetön is composed of thick *Sphagnum* peat moss, and is very steep except in two small places where *Fontinalis antipyretica* Hedw. grows on the bottom. In Horkkajärvi the mineral soil reaches the shoreline, although the shore is steep. The only floating-leaved macrophyte is *Nuphar lutea* (L.) Smith, which is scarce in Horkkajärvi.

Lake Nimetön has two small inlets and one outlet. Lake Horkkajärvi has two inlets, which bring water from a clearcutting area, and one outlet. The primary rock in this area is granodiorite and gneiss (geological map of Finland, sheet-2134-Lammi). The soil is morainic and the forest is mainly composed of scots pine and spruce.

Hydrobiologia 101, 105–110 (1983).

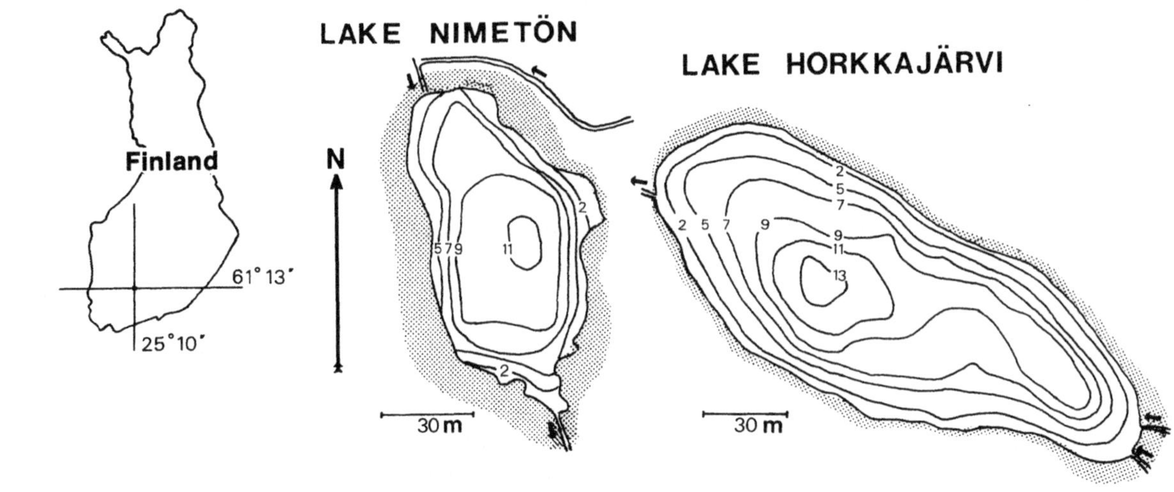

Fig. 1. Location of the study area and the bathymetry of the lakes.

Table 1. Some morphometric and physico-chemical properties of the study lakes. Physico-chemical data represent mainly the average summer situation.

	Lake Nimetön		Lake Horkkajärvi	
	m 0–1	m 6	m 0–1	m 6
Surface area (ha)	0.4		1.1	
Volume (10^3 m^3)	33.6		78.6	
Max. depth (m)	11		13	
Mean depth (m)	8		7	
Drainage basin (ha)	34		70	
pH	5.0–6.0	5.5–6.0	4.6–6.0	5.8–6.2
Colour (mg Pt · l^{-1})	120–200	200–300	200–300	350–400
Conductivity (μS · cm^{-1} 20 °C)	40	50–60	30	50–60
Total N (μg · l^{-1})	300–700	900–2 200	500–700	900–2 000
$NO_3 + NO_2$ (μg · l^{-1})	<15	<20	<15	<25
NH_4 (μg · l^{-1})	<2	40–100	<2	40–80
Total P (μg · l^{-1})	10–40	150–250	20–30	150–200
PO_4 (μg · l^{-1})	<5	50–230	<3	80–200

Ice cover on the lakes lasts for six months, from November to May.

Materials and methods

Phytoplankton samples were taken weekly in summer (May–October) and monthly in winter. Samples were taken with a Ruttner sampler from the following depths: 0, 0.5, 1, 2, 3, 5 and 7 m. Primary production *(in situ)* was measured in summer every week from the depths 0.1, 0.5, 1 and 2 m.

Chlorophyll *a* was determined with methanol according to Holm-Hansen & Riemann (1978), except at the beginning of the summer when the analyses were made with methanol according to Golterman *et al.* (1978).

The phytoplankton species composition and biomass were determined using an inverted phase contrast microscope according to Utermöhl (1958). The sampels were preserved in acid Lugol's solution.

Primary production of phytoplankton was measured by ^{14}C method in 130 ml glass bottles during 24 hours. The samples were filtered with Millipore HAWP 29325, 0.45 μm membrane filters with a pressure less than 0.3 kPa · cm^{-2}. Radioactivity was assayed with a Geiger-Müller beta counter. Dissolved inorganic carbon for the measurement of primary production was determined with a carbon

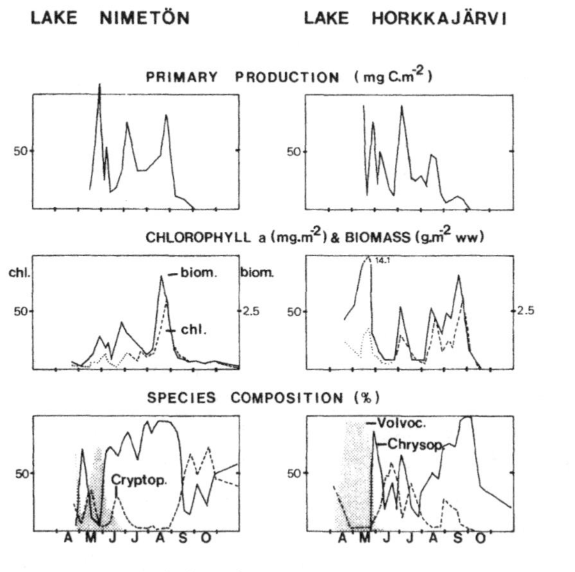

Fig. 2. Seasonal variation of primary production, chlorophyll *a*, phytoplankton biomass and species composition (% of total biomass) in Lake Nimetön and Lake Horkkajärvi.

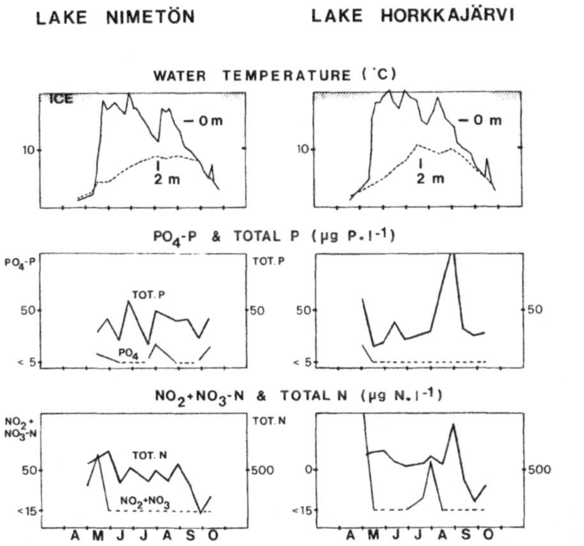

Fig. 3. Seasonal variation of water temperature, dissolved inorganic nutrient (PO_4 and $NO_2 + NO_3$) and total nitrient (P and N) concentrations.

analyzer (Salonen 1981).

The nutrient analyses (PO_4-phosphorus and NO_2- and NO_3-nitrogen) were made with an AKEA automatic chemical analysis system, using the molybdate acid method introduced by Murphy & Riley (1962; see also Grasshoff 1976) for phosphate,

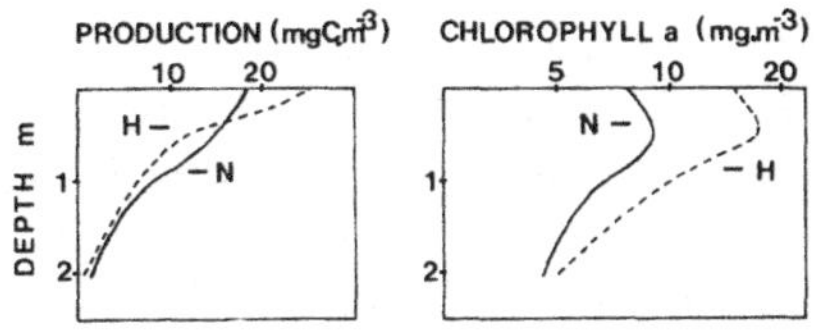

Fig. 4. The mean vertical distribution of primary production and chlorophyll *a* in epilimnion during the growing season in Lake Nimetön and Lake Horkkajärvi.

and the method of Wood *et al.* (1967) for nitrite and nitrate, reducing nitrate to nitrite in cadmium-copper columns. Total nitrogen and total phosphorus were also determined with the AKEA system after boiling the samples in an autoclave for 30 min with peroxidisulphate in alkaline solution (Koroleff 1979).

Results

Remarkable seasonal variations were observed in primary production in these lakes during 1979 (Fig. 2). In spring when snow was melting and solar radiation began to penetrate the ice, primary production increased rapidly. After the ice melted at the beginning of May (12.5.) the surface waters warmed very quickly (Fig. 3) and the production increased greatly. The spring maximum was noted about two weeks after the ice had melted in both lakes. At that time the water temperature was ca. +18 °C on the surface of the lakes.

After the spring maximum there was a rapid decrease in primary production in June. Later two summer maxima occurred at the beginning of July and in the second half of August. The production was stopped at the beginning of September in Lake Nimetön and at the end of the month in Lake Horkkajärvi.

The mean daily production at the surface of Nimetön was 36.8 mg C · m^{-3} and in Horkkajärvi 50.3 mg C · m^{-3} (May–September). At a depth of one meter the values were about 50% of the surface values in Nimetön and 30% in Horkkajärvi (Fig. 4). Owing to the thin productive water layer and short growing period (ca. 160 days) the annual production was quite small: in Nimetön 6.0 g C · m^{-2} and in Horkkajärvi 5.6 g C · m^{-2}.

Also the seasonal variation of chlorophyll *a* and biomass were distinct in both lakes (Fig. 2): phytoplankton had three maxima in Lake Nimetön and three to four maxima in Horkkajärvi. During the maxima the concentration of chlorophyll *a* could reach 60 mg · m^{-2} in epilimnion, but during the minima it was only 5–10 mg · m^{-2}. The seasonal patterns of biomass and chlorophyll *a* were very similar except at the beginning of the summer when chlorophyll *a* was determined according to Golterman *et al.* (1978). This method gave too low concentrations because of the heat treatment and too strong acidification (e.g., Arvola 1981). Therefore the high spring maximum of phytoplankton (e.g., in Horkkajärvi biomass was 14.1 g · m^{-2} ww) was not clearly seen in chlorophyll (cf. Fig. 2). Excluding the beginning of the summer correlation coefficient between chlorophyll *a* and biomass was 0.91^{XXX} ($p < 0.001$) in Lake Nimetön and 0.88^{XXX} ($p < 0.001$) in Lake Horkkajärvi. However, chlorophyll concentration and primary production had no interrelationship in these lakes. The mean biomass of the uppermost two meters in Nimetön was 1.1 g · m^{-2} and in Horkkajärvi 2.2 g · m^{-2}. The biomass was concentrated to the uppermost surface layer, then the biomass as well as the primary production decreased rapidly below the 0.5 m depth (Fig. 4). The concentrations of dissolved inorganic phosphorus (PO_4) and nitrogen ($NO_2 + NO_3$) were usually below the detection limit of the methods (Fig. 3). Therefore it is meaningless to calculate any correlations between nutrients and primary production.

The phytoplankton species composition was very similar in both lakes (Fig. 2). During the spring maximum Volvocales, especially *Chlamydomonas* species were dominant. The community was composed almost purely of a *Chlamydomonas* green alga species, with a diameter of ca. 15–20 µm. In summer Chrysophyceae were very common; *Uroglena* species in early summer and *Mallomonas caudata* later. Cryptophyceae were quite common

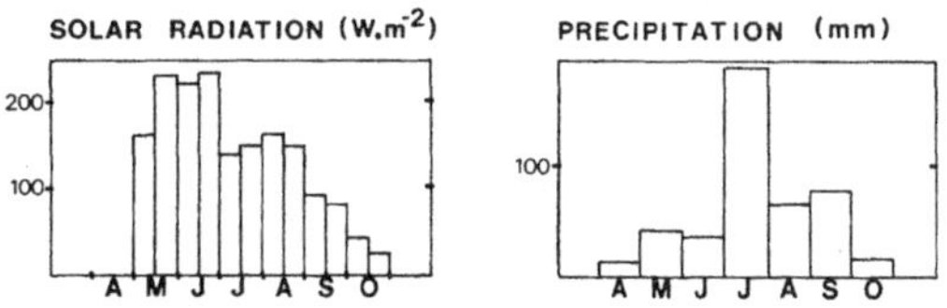

Fig. 5. The variation of main daily solar radiation and monthly precipitation at Lammi Biological Station, 20 km SW from the study area.

too but Diatomae and Cyanophyceae were very rare.

During winter Chlorophyceae, Chrysophyceae and Cryptophyceae were dominant. The number of species during that time was extremely low, and winter biomasses were very small, usually below 0.01 g · m^{-2}.

Discussion

The growing season of phytoplankton in the lakes under study began already before the lakes were ice-free, because light, which is the minimum factor during winter, could penetrate the ice as soon as snow melted. *Chlamydomonas* green algae could obviously best utilize the dark and cold environment which dominates at that time and therefore this group grew very strongly in spring (cf. Ramberg 1979). This, like the vernal peaks, have also been noticed in some other northern lakes (Willen 1962; Janus & Dutchie 1979; Ramberg 1979; Arvola 1980). Possibly *Chlamydomonas,* which has two flagella, is able to migrate vertically so that it can optimize its location in relation to light and nutrients. If this is the situation, the absence of water circulation after the ice has melted, can favour flagellated species, such as *Chlamydomonas.* Alleiopathic interactions may also help this species to support its monoculture (cf. Harris 1971; McCracken *et al.* 1980). The decline of this community was obviously a result of the depletion of nutrients (cf. Fig. 4).

After the spring maximum, dissolved inorganic nutrient concentrations were very low, and in June *Uroglena* and *Dinobryon* species were the most common in phytoplankton community. According to Rodhe (1948) they have very low phosphorus requirements, which seems to be in accordance with nutrient loss in the lakes, although the relationship between nutrients and phytoplankton is unclear owing to the low inorganic nutrient contents.

The summer maximum began to develop already at the beginning of July, but the cold and rainy weather prevented this development and the maximum could arise only after this period (Figs. 2 and 5). The summer maximum was mainly composed of *Mallomonas caudata* in both lakes. When this community broke down in September, primary production, biomass and chlorophyll content decreased quickly. At the same time in September also the total nutrient concentrations as well as solar radiation and water temperature decreased (cf. Figs. 3 and 5). On the other hand, throughflow of water increased during September as a result of precipitations which caused great biomass loss. The uppermost one meter water layer can change in these lakes totally during three to five days in consequence of strong precipitations.

During the period of growth various factors control phytoplankton succession and primary production in the lakes studied. Light, nutrients, temperature and flushing seem to be the main factors which affect phytoplankton community and productivity. Another serious limitation for primary production in these polyhumic lakes is the dark-brown water. The thickness of the euphotic zone is only one to two meters. Probably the phytoplankton composition, almost all flagellated species, is determined by the light regime in these lakes (cf. Ramberg 1979; Ilmavirta 1980).

The interrelationships between phytoplankton and zooplankton are unknown but the relation between primary production and total respiration of plankton is better known (cf. Salonen *et al.* 1983). It seems that phytoplankton produces less organic matter for the ecosystem than plankton consumes during the same time, because primary production was in 1980 only about half the total respiration in the epilimnion. This seems to be a paradox, but it must be remembered that the measured primary production values of this study are too low, and the extracellular production was not measured (cf. Chróst *et al.* 1981) and the water of these lakes has a very high humus concentration, which is a potential energy resource for plankton.

The productivity of the study lakes is of the same level as in an oligotrophic lake (Lake Pääjärvi, Ilmavirta & Kotimaa 1974) near the study area, but noticeably lower than in an eutrophic lake (Lake Lovojärvi, Keskitalo 1976, 1977; Heikkola 1980), also in the vicinity of the study area. Compared with the small forest lakes of Central Sweden (Ramberg 1979) the productivity of Lake Nimetön and Lake Horkkajärvi was about at the same level or a little greater.

According to these preliminary results humus together with some morphometric and physicochemical properties is one of the main factors, which affect many kinds of biological processes in small, polyhumic lakes. Our knowledge of the effects of, e.g., humus compounds on primary production or the interrelationship between humus and nutrients is defectively known (cf. Jackson & Hecky 1980), as are many questions pertinent to the effects of humus. These questions, however, are important when trying to understand the function of lake ecosystems in northern coniferous areas, because humus contents are usually high in the lakes of this region. In this respect small forest lakes offer very interesting 'laboratories' to research.

Acknowledgements

I would like to thank K. Salonen and M. Rask for their help in the field and for their valuable comments on the manuscript as well as the other persons which have contributed to advance of this work. I also thank R. Jones and M. Nikinmaa, who checked the English of the manuscript. This study was supported by the Maj and Tor Nessling Foundation.

References

Arvola, L., 1980. On the species composition and biomass of the phytoplankton in the Lokka reservoir, northern Finland. Ann. Bot. Fennici, 17: 325-335.

Arvola, L., 1981. Spectrophotometric determination of chlorophyll *a* and phaeopigments in ethanol extractions. Ann. Bot. Fenn. 18: 221-227.

Chróst, R. J., Siuda, W. & Wazyk, M., 1981. Primary production, extracellular release of organic matter by phytoplankton and heterotrophic activity of bacteria in three lakes in the Mazurian Lake District, Poland. Pol. Arch. Hydrobiol. 28: 11-18.

Golterman, H. L., Clymo, R. S. & Ohnstadt, M. A. M., 1978. Methods of physical and chemical analysis of freshwaters. IBP Handbook 8, 214 pp. Blackwell Sci. Publ., Oxford, London.

Grasshoff, K., 1976. Procedures for the automatic determination of seawater constituents. In: Grasshoff, K. (ed.). Methods of Seawater Analysis, pp. 276-289. Verlag Chemie, Weinheim.

Harris, D. O., 1971. Growth inhibitors produced by the green algae Volvocaceae. Arch. Mikrobiol. 76: 47-50.

Heikkola, K., 1980. Lovojärven kasviplanktonin tuotannosta ja siihen vaikuttavista tekijöistä (On the primary production of phytoplankton in Lake Lovojärvi and the factors affecting it). Laudatur study. Dep. Bot., Helsinki Univ., 32 pp.

Holm-Hansen, O. & Riemann, B., 1978. Chlorophyll *a* determination: improvements in methodology. Oikos 30: 438-447.

Ilmavirta, K. & Kotimaa, A-L., 1974. Spatial and seasonal variations in phytoplanktonic primary production and biomass in the oligotrophic lake Pääjärvi, southern Finland. Ann. Bot. Fenn. 11: 112–120.

Ilmavirta, V., 1980. Phytoplankton in 35 finnish brown-water lakes of different trophic status. In: Dokulil, M., Metz, H. & Jewson, D. (eds.). Dev. hydrobiol., pp. 121–130, Junk, The Hague.

Jackson, T. A. & Hecky, R. E., 1980. Depression of primary productivity by humic matter in lake and reservoir waters of the boreal forest zone. Can. J. Fish. aquat. Sci. 37: 2300–2317.

Janus, L. L. & Dutchie, H. C., 1979. Phytoplankton composition and periodicity in a northeastern Quebec lake. Hydrobiologia 63: 129–134.

Keskitalo, J., 1976. Phytoplankton pigment concentrations in the eutrophicated Lake Lovojärvi, south Finland. Ann. Bot. Fenn. 13: 27–34.

Keskitalo, J., 1977. The species composition and biomass of phytoplankton in eutrophic Lake Lovojärvi, southern Finland. Ann. Bot. Fenn. 14: 71–81.

Koroleff, F., 1979. Methods for the chemical analysis of seawater. Meri 7: 1–60 (in Finnish).

Lehmusluoto, P. O. & Ryhänen, R., 1972. Lake Hakojärvi, a polyhumic lake in southern Finland. Verh. int. Ver. Limnol. 18: 403–408.

McCracken, M. D., Middaugh, R. E. & Middaugh, R. S., 1979. A chemical characterization of an algal inhibitor obtained from Chlamydomonas. Hydrobiologia 70: 271–276.

Meriläinen, J., 1970. On the limnology of the meromictic Lake Valkiajärvi, in the Finnish Lake District. Ann. Bot. Fenn. 7: 29–51.

Murphy, J. & Riley, J. P., 1962. A modified single-solution method for the determination of phosphate in natural waters. Anal. chim. Acta 27: 31–36.

Ramberg, L., 1979. Relations between phytoplankton and light climate in two Swedish forest lakes. Int. Revue ges. Hydrobiol. 64: 749–782.

Rodhe, W., 1948. Environmental requirements of freshwater pankton algae. Symb. bot. ups. 10: 1–149.

Salonen, K., 1981. Rapid and precise determination of total inorganic carbon and some gases in aqueous solutions. Wat. Res. 15: 403–406.

Salonen, K., Kononen, K. & Arvola, L., 1983. Total respiration of plankton in two small polyhumic lakes. Hydrobiologia, this volume.

Utermöhl, H., 1958. Zur Vervollkommung der quantitativen Phytoplanktonmethodik. Mitt. int. Ver. Limnol. 9: 1–38.

Willen, T., 1962. Studies on the phytoplankton of some lakes connected with or recently isolated from the Baltic. Oikos 13: 167–199.

Wood, E. D., Armstrong, F. A. J. & Richards, F. A., 1967. Determination of nitrate in sea water by cadmium-copper reduction to nitrite. J. mar. biol. Ass. U.K. 47: 23–31.

Zooplanktic indicators of trophy and their food

Birger Pejler
Institute of Limnology, Box 557, S-751 22 Uppsala, Sweden

Keywords: forest water ecosystems, zooplankton, feeding, indicators, hypertrophy, eutrophy, oligotrophy, ultraoligotrophy

Abstract

Regarding specific food requirements zooplankters in oligotrophic forest lakes are compared with indicators of other trophic conditions.

Most zooplanktic indicators of eutrophic conditions feed on very small particles, to a large extent on bacteria, while the indicators of oligotrophy may be filtrators of courser particles. Both categories comprise raptors as well. In both extremes of lake types, hypertrophy and ultraoligotrophy, feeders on minute particles increase in relative importance, while moderate eutrophy or oligotrophy render a higher diversity of species and feeding categories.

Introduction

When trying to characterize the specific food requirements of zooplankters in oligotrophic forest lakes, one approach is to focus the interest on its indicative species and to compare these with indicators of other trophic conditions. For this reason an analysis, based on the literature, was performed concerning the food of oligotrophic as well as eutrophic indicators.

Pertinent problems of this type were treated by Gliwicz (1969, 1974, 1977, 1980), Hillbricht-Ilkowska & Weglеńska (1970), Hillbricht-Ilkowska (1977), Nilssen (1978) and Geller & Müller (1981). The feeding conditions of the dominant zooplankters, especially those of crustaceans, were discussed, and comparisons were made between different lakes as well as between different seasons within the same lake. In the present paper the results have been complemented, especially by data on rotifers, previously not discussed in this connection. The attention is drawn to species regarded as indicators of eutrophic or oligotrophic conditions, mainly according to Pejler (1965). The Swedish lakes in question represent a much broader spectrum of the trophic continuum than the Polish lakes investigated by the above authors.

Food of oligotrophic indicators

There are strong reasons to include *Asplanchna herricki* de Guerne (see Pejler 1957b; Hakkari 1972) and *Cyclops scutifer* Sars (cf. Elgmork 1967; Kiefer 1978) with the species enumerated by Pejler (1965, p. 503).

Three of the rotifers indicative of oligotrophy may be designated as 'raptorial forms' (according to the terminology in Hutchinson 1967; called 'Greifer' by Naumann 1923; 'snatchers' by Pejler 1965, p. 477; 'seizers' by Nilssen 1976; 'grasping species' by Pourriot 1977), viz. *Asplanchna herricki* (see Monakov 1972; Gilbert 1980), *Synchaeta grandis* Zacharias (Koste 1978) and *Ploesoma hudsoni* (Imhof) (see Dumont 1977; Pourriot 1977; Guiset 1977).

Among the cladocerans *Holopedium gibberum* Zaddach has a varying and broad filtering range,

Hydrobiologia 101, 111–114 (1983).

with a probable filtering optimum at a larger particle size than in *Bosmina* and *Daphnia* (Johansson *et al.* 1976). This assumption is corroborated in the SEM studies performed by Geller & Müller (1981), who consider its range of high filtering efficiency to reach from 4 to at least 15, and perhaps more than 20 µm. Regarding the genus *Daphnia* it has been shown that different species can utilize different food as regards quality (Burns 1968; Infante 1973), and Geller & Müller (1981) have found there are differing ranges of high filtering efficiency, that of *D. galeata* Sars reaching from about 1 to 20 µm.

According to Naumann (1923) and Monakov (1972) the copepod *Heterocope appendiculata* Sars feeds at least in the adult stage as a raptor, which is considered to be the case in all cyclopoids as well (Fryer 1957). This means that both animals and bigger algae can be utilized. The omnivorous character of *H. appendiculata* and *Cyclops scutifer* is stressed by Langeland & Reinertsen (1982) as well, basing themselves on own analyses of stomach contents. Regarding *C. scutifer* the following circumstances are of interest in this connection: in lake Kuoblatjåkkojaure, which is probably the highest situated lake in all Sweden, only two zooplankters were found, viz. the rotifer *Keratella hiemalis* Carlin and the copepod *Cyclops scutifer* (Pejler 1962). In this lake and in similar lakes the most important groups of the phytoplankton are chrysomonads and cryptomonads, often also dinoflagellates (Rodhe 1962; Rodhe *et al.* 1966). To a large extent they consist of minute forms ('µ-algae', with a size of 1–2 µ or less). Though *C. scutifer* is a potential predator it cannot feed extensively on other zooplankters, especially as it is often the dominating zooplankter (see Rodhe 1955; Table 1). For similar reasons cannibalism probably does not play an important role, and the copepod has to be considered as mainly herbivourous. In fact Taube & Nauwerck (1967) and Gerritsen (1978) have bred this species on exclusively algal food.

Food of eutrophic indicators

Some species enumerated by Pejler (1965, pp. 502–503, under point 5) are here discussed. In addition, a *pelagical* occurrence of *Chydorus sphaericus* Müll. ought to be considered indicative of eutrophy (op. cit., p. 470).

Table 1. Division of zooplanktic indicators into different feeding categories. See further the text. Only those species are included for which the feeding is known in a satisfying way. The three filtrator categories are taken from Geller & Müller (1981).

	(1)	(2)	(3)	(4)	(5)
Indicators of oligotrophy					
Asplanchna herricki					*
Synchaeta grandis					*
Ploesoma hudsoni					*
Holopedium gibberum				*	
Daphnia galeata			*		
Heterocope appendiculata					*
Cyclops scutifer					*
Indicators of eutrophy					
Brachionus angularis	*				
Brachionus calyciflorus	*				
Anuraeopsis fissa	*				
Polyarthra euryptera					*
Pompholyx sulcata	*				
Filinia longiseta	*				
Daphnia cucullata		*			
Chydorus sphaericus		*			

(1) Sedimentators.
(2) High-efficiency bacteria feeders.
(3) Low-efficiency bacteria feeders.
(4) Macrofiltrators.
(5) Raptorial forms.

The species of *Brachionus* are comparatively well known in this respect (see especially the reviews by Pourriot in 1977 and by Starkweather in 1980). The food is quite varying, consisting of diverse types of algae as well as of detritus and bacteria. The last-named item seems to be especially important for some species, e.g. *B. angularis* Gosse. Also *B. calyciflorus* Pallas, previously considered to feed mainly on algae, was cultured on the bacterium *Aerobacter aerogenes* for more than 40 generations, with no decrease in net reproductive rate (Starkweather *et al.* 1979). Furthermore, *Anuraeopsis fissa* (Gosse) as well as *Pompholyx sulcata* Hudson and *Filinia longiseta* (Ehrbg) feed predominantly on detritus and bacteria (Pourriot 1977). Finally, *Polyarthra euryptera* (Wierzejski) belongs to the raptors (see Pourriot 1977) and can feed on relatively large algae like *Cryptomonas ovata* Ehrbg (Dieffenbach & Sachse 1911) and *Peridinium* (Pejler 1961; Starkweather 1980).

Daphnia cucullata Sars feeds, on an average, on significantly smaller particles than *D. galeata* (Geller & Müller 1981). Its optimal filtering range is from below 0.5 to about 5 µm, and it is by the mentioned authors included in the category 'high efficiency bacteria feeders', in contrast to *D. galeata* which is considered a 'low efficiency bacteria feeder'. *Chydorus sphaericus,* with its optimal filtering range of 0.4–ca. 2 μm, belongs to the first-mentioned group as well (op. cit.). Also Gliwicz (1969) found its filtering range to be unusually small compared to other cladocerans, though he based his findings on another method than Geller and Müller.

Discussion

In connection with an increasing trophic degree, specialists on small-particle feeding are favoured and macrofiltrators disfavoured. Consequently, a successively larger fraction of the food will consist of bacteria. The survivors in hypertrophic lakes thus mostly consist of high-efficiency bacteria feeders and of raptors, which also seems to be in accordance with Patalas (1954), Pejler (1965) and Andersson *et al.* (1975). This can be achieved partly by a change in species composition, partly by changing proportions in the diet, implying that a greater fraction will consist of bacteria for the sedimentators and microfiltrators.

A change in the food uptake, connected with different trophic degrees, was observed also in a raptor, *Asplanchna priodonta* Gosse, by Ejsmont-Karabin (1974), the animal fraction of the food being much greater in the eutrophic than in the mesotrophic lake investigated. The present author studied, together with Hans Olofsson, the stomach contents of this species in still more oligotrophic lakes of central Sweden, whereby no animal remains were found at all. The visible contents consisted exclusively of larger algae such as peridineans.

In the extremely poor, i.e. ultraoligotrophic lakes, the number of steps of the ecological pyramid should be relatively small, and all zooplankters could be conceived to be herbivorous. It is interesting to note that in the rotifer community only suspension feeders (i.e. non-predatory forms) were found in all investigated arctic lakes of northern Sweden (Pejler 1957a, Table 29). Most subarctic lakes contained two or more such species, in one case (Lake Nedre Laksjön) as many as 5.

In other extreme arctic environments we meet a similar picture. Thus, in the poor large lakes of Spitsbergen Amrén (1964) found only suspension-feeding rotifers and *Cyclops,* though a species of *Daphnia* occurred in richer ponds. Similar conditions prevail on the island of Bjørnøya (Fleetwood *et al.* 1974).

The occurrence of an unusually high share of nannoplanktic algae in ultraoligotrophic lakes is stressed, e.g., by Pavoni (1963) and Gliwicz (1975) as well. Similarly in the lakes on Spitsbergen investigated by Amrén, small flagellates played a very important role (Willén 1980). It is astonishing to find very small particles (μ-algae) to constitute an important food source in ultraoligotrophic arctic lakes.

Along the trophic scale, beginning at the ultraoligotrophic extreme, the number of planktic rotifer species successively increases up to a certain point (roughly in the 'mesotrophic' area), after which the species number declines again towards the hypertrophic end (Pejler 1957c). The same most likely also occurs if the whole zooplanktic community is considered. At the same time the number of functional groups both increases and decreases. The category most typical at both extremes is the small-particle feeders, whereby the food is predominated by μ-algae or bacteria. Obviously raptors appear as well. When approaching the center of the scale other functional groups increase in importance, such as macrofiltrators, 'low-efficiency bacteria feeders' and suckers.

References

Amrén, H., 1964. Ecological and taxonomical studies on zooplankton from Spitsbergen. Zool. Bidr. Upps. 36: 209–276.

Andersson, G., Berggren, H. & Hamrin, S., 1975. Lake Trummen restoration project. III. Zooplankton, macrobenthos and fish. Verh. int. Ver. Limnol. 19: 1097–1106.

Burns, C. W., 1968. Direct observations of mechanisms regulating feeding behavior of Daphnia, in lakewater. Int. Revue ges. Hydrobiol. Hydrogr. 53: 83–100.

Dieffenbach, H. & Sachse, R., 1912. Biologische Untersuchungen an Rädertieren in Teichgewässern. Int. Revue ges. Hydrobiol. Hydrogr. 3: 1–93.

Dumont, H. J., 1977. Biotic factors in the population dynamics of rotifers. Arch. Hydrobiol. Beih. 8: 98–122.

Ejsmont-Karabin, J., 1974. Research on the feeding of planktonic polyphage Asplanchna pirodonta Gosse (Rotatoria). Ekol. pol. A 22: 311–317.

Elgmork, K., 1967. On the distribution and ecology of Cyclops scutifer Sars in New England (Copepoda, Crustacea). Ecology 48: 967–971.

Fleetwood, Å., Pejler, B., Einsle, U. & Arnemo, R., 1974. Stratigraphical and biological investigations of some Bjørnøya lakes. Norsk Polarinst. Årbok 1972: 63–71.

Fryer, G., 1957. The feeding mechanism of some freshwater cyclopoid copepods. Proc. zool. Soc. Lond. 129: 1–25.

Geller, W. & Müller, H., 1981. The filtration apparatus of Cladocera: Filter mesh-sizes and their implications on food selectivity. Oecologia 49: 316–321.

Gerritsen, J., 1978. Instar-specific swimming patterns and predation of planktonic copepods. Verh. int. Ver. Limnol. 20: 2531–2536.

Gilbert, J. J., 1980. Feeding in the rotifer Asplanchna: Behavior, cannibalism, selectivity, prey defenses, and impact on rotifer communities. In: Kerfoot, W. C. (ed.), Evolution and Ecology of Zooplankton Communities, pp. 158–172, University Press of New England.

Gliwicz, Z. M., 1969. Studies on the feeding of pelagic zooplankton in lakes with varying trophy. Ekol. pol. A 17: 663–708.

Gliwicz, Z. M., 1974. Trophic status of freshwater zooplankton species. Wiad. ekol. 20: 197–206 (Polish, with English summary).

Gliwicz, Z. M., 1975. Effect of zooplankton grazing on photosynthetic activity and composition of phytoplankton. Verh. int. Ver. Limnol. 19: 1490–1497.

Gliwicz, Z. M., 1977. Food size selection and seasonal succession of filter feeding zooplankton in a eutrophic lake. Ekol. pol. 25: 179–225.

Gliwicz, Z. M., 1980. Filtering rates, food size selection, and feeding rates in cladocerans – Another aspect of interspecific competition in filter-feeding zooplankton. In: Kerfoot, W. C. (ed.), Evolution and ecology of zooplankton communities, pp. 282–291, University Press of New England.

Guiset, A., 1977. Stomach contents in Asplanchna and Ploesoma. Arch. Hydrobiol. Beih. 8: 126–129.

Hakkari, L., 1972. Zooplankton species as indicators of environment. Aqua fenn. pp. 46–54.

Hillbricht-Ilkowska, A. & Wegleńska, T., 1970. Some relations between production and zooplankton structure of two lakes of a varying trophy. Pol. Arch. Hydrobiol. 17: 233–240.

Hillbricht-Ilkowska, A., 1977. Trophic relations and energy flow in pelagic plankton. Pol. ecol. Stud. 3: 3–98.

Hutchinson, G. E., 1967. A Treatise on Limnology. 2. Introduction to lake biology and the limnoplankton. John Wiley & Sons, Inc., New York, 1115 pp.

Infante, A., 1973. Investigations on the incorporation of different algae by some zooplankton-species. Arch. Hydrobiol. Suppl. 42: 340–405 (German, with English summary).

Johansson, J.-Å., Olofsson, H. & Ramberg, L., 1976. Studier av zooplanktons konsumtion i Botjärn. Klotenproj. rapport 5. Uppsala univ. limnol. inst., 27 pp. (Swedish).

Kiefer, F., 1978. Freilebende Copepoda. In: Elster, H.-J. & Ohle, W. (eds.). Die Binnengewässer, bd 26: Das Zooplankton der Binnengewässer. II, pp. 1–343,

Koste, W., 1978. Das Rädertier-Porträt: Synchaeta grandis, ein in Mitteleuropa vom Aussterben bedrohtes Planktonrädertier. Mikrokosmos pp. 331–336.

Langeland, A. & Reinertsen, H., 1982. Interactions between phytoplankton and zooplankton in a fertilized lake. Holarct. Ecol. 5: 000–000.

Monakov, A. V., 1972. Review of studies on feeding of aquatic invertebrates conducted at the institute of biology of inland waters, Academy of Science, USSR. J. Fish. Res. Bd Can. 29: 363–383.

Naumann, E., 1923. Spezielle Untersuchungen über die Ernährungsbiologie des tierischen Limnoplanktons. II. Acta Univ. lund., nov. ser., avd. 2, bd 19, 6, 17 pp.

Nilssen, J. P., 1976. Community analysis and altitudinal distribution of limnetic Entomostraca from different areas in southern Norway. Pol. Arch. Hydrobiol. 23: 105–122.

Nilssen, J. P., 1978. Eutrophication, minute algae and inefficient grazers. Mem. Ist. ital. Idrobiol. 36: 121–138.

Patalas, K., 1954. Pelagic crustacean complexes of 28 Pommeranian lakes. Ekol. pol. 2: 61–92. (Polish, with English summary).

Pavoni, M., 1963. Die Bedeutung des Nannoplanktons im Vergleich zum Netzplankton. Schweiz Z. Hydrol. 25: 219–341.

Pejler, B., 1957a. Taxonomical and ecological studies on planktonic Rotatoria from northern Swedish Lapland. K. svenska VetenskAkad. Handl., ser. 4, bd 6, 5, 68 pp.

Pejler, B., 1957b. Taxonomic and ecological studies on planktonic Rotatoria from Central Sweden. K. svenska Vetensk-Akad. Handl., Ser. 4, bd 6, 7, 52 pp.

Pejler, B., 1957c. Studies on the taxonomy and ecology of planktonic Rotatoria. Thesis, Uppsala, 11 pp.

Pejler, B., 1961. The zooplankton of Ösbysjön. Djursholm. I. Seasonal and vertical distribution of the species. Oikos 12: 225–248.

Pejler, B., 1962. Taxonomic notes on some planktic freshwater rotifers. Zool. Bidr. Upps. 35: 307–319.

Pejler, B., 1965. Regional-ecological studies of Swedish freshwater zooplankton. Zool. Bidr. Upps. 36: 407–515.

Pourriot, R., 1977. Food and feeding habits of Rotifera. Arch. Hydrobiol. Beih. 8: 243–260.

Rodhe, W., 1955. Can plankton production proceed during winter darkness in subarctic lakes? Verh. int. Ver. Limnol. 12: 117–122.

Rodhe, W., 1962. On phytoplankton production in high mountain transparent lakes. Mem. Ist. ital. Idrobiol. 15: 21–28. (Italian, with English summary).

Rodhe, W., Hobbie, J. E. & Wright, R. T., 1966. Phototrophy and heterotrophy in high mountain lakes. Verh. int. Ver. Limnol. 16: 302–313.

Starkweather, P. L., 1980. Aspects of the feeding behavior and trophic ecology of suspension-feeding rotifers. Hydrobiologia 73: 63–72.

Starkweather, P. L., Gilbert, J. J. & Frost, T. M., 1979. Bacterial feeding by the rotifer Brachionus calyciflorus: Clearance and ingestion rates, behavior and population dynamics. Oecologia 44: 26–30.

Taube, I. & Nauwerck, A., 1967. Zur Populationsdynamik von Cyclops scutifer Sars. I. Rep. Inst. Freshwat. Res. Drottningholm 47: 76–86.

Willén, T., 1980. Phytoplankton from lakes and ponds on Vestspitsbergen. Acta phytogeogr. suec. 68: 173–188.

G a (bryozoa) in the outlet of a humic acid lake

G
1 logy, Univ. of Bergen, N-5014 Bergen, Norway
2 ?n, Norway

K istory, growth and nutrition

Abstract

Monthly observations on the bryozoa *Fredericella sultana* have been carried out in the outlet of a humic acid lake for more than two years. The growth of the colonies were rapid during algal bloom and in fall the colonies almost covered the free underside of nearly all stones on the bottom.

In winter most of the colonies died out, but scattered individuals were also observed alive during this period. Statoblasts produced in summer were 'hatching' in March–April next year.

F. sultana seems to feed on all particles that can pass through the mouth opening. A special study regarding this was made on the diatoms, by analysis of fecal pellets. It was found that young colonies of *F. sultana* filtrated most efficiently the smallest particles (<20 μm) while older individuals in addition were able to elevate the filtration of bigger particles (40–100 μm). The feeding takes place both by filtration by the cilia and by an active use of the tentacles themselves on the lophophore.

Introduction

Freshwater bryozoans, the moss animals, are sessile, filter-feeding and colony-building organisms. The reproduction at our latitude is vegetative by budding and production of statoblasts.

In Norway 7 species of moss animals are recorded. A research for the animals in 57 lakes/rivers in western Norway showed that 38 of the localities were holding bryozoans (Raddum 1973). One of the most distributed species was *Fredericella sultana* (Blumenbach) inhabiting high mountain oligotrophic localities as well as eutrophic ones in the lowland. The species was most frequent near the shoreline on substratum like leaves of Numphaea, roots, stones or other objects present in the water. Individuals also occurred in Ekman samples taken from different depths down to 20 m in oligotrophic lakes. This depht is, however, no limit as records of the species elsewhere have been made both from 65 and 214 m depth (Heding 1938; Marcus 1940).

In running water, especially near outlets of lakes, *F. sultana* may develop dense colonies on the underside of stones caused by the high concentration of nutrition particles in such habitats. In this paper the growth, life history and nutrition of *F. sultana* will be reported from the outlet of a little bog lake (Ulsetstemma) near Bergen.

Material and methods

In the period July 1970–October 1972 sampling and observations of *F. sultana* were carried out every second week during the growing season and once a month during winter.

Five quantitative samples (one sample = the colony mass covering 3 cm^2 stone surface) were taken five times between April and October 1972. Each time the samples were put together and preserved in 75% alcohol. Biomass is given as dry weight after one hour at 105 °C.

Hydrobiologia 101, 115–120 (1983).

Germination of statoblasts and development of colonies were studied in small aquaria and at the locality during the whole research period.

The nutrition of *F. sultana* was investigated between April and July 1981. Samples of phytoplankton were taken near colonies of *F. sultana* on 250 ml brown bottles and fixed with acid Lugol solution. Counting of algae was carried out on an inverted microscope after Uthermöhls method (1931, 1958). Samples of benthic diatoms were taken near the colonies of *F. sultana* and fixed with neutralized formaldehyde. The diatoms were put in concentrated $KMnO_4$ for one day and then heated in concentrated HCl and washed with distilled water before mounting in Hyrax. At least 300 valves were identified from each sample.

The organisms eaten by *F. sultana* were investigated by analysis of fecal pellets produced by colonies kept one day in aquaria holding water from the locality. One half of the pellets were fixed in neutralized formaldehyde for diatom analysis. The other half was put in 30 ml distilled water and 1 ml 20% neutralized formaldehyde and broken by shaking for 3 min. Counting of algae was carried out as for phytoplankton. Hustedt (1930) and Prescott (1970) were mostly used for identification of the algae.

Water samples for chemical analyses were regularly taken both in 1970–1972 and in 1981.

Results and discussion

Water chemistry

The ionic content and pH found in Lake Ulsetstemma was typical for lakes situated in forest areas in Norway (Table 1).

The water was humic with colour ranging between 25 and 143 mg Pt · l^{-1}, pH between 5.0–6.0, conductivity between 25–60 $\mu S \cdot cm^{-1}$ and Ca in the range of 0.8–2.1 mg · l^{-1}.

The content of particulated organic matter (POM) was usually <5 mg · l^{-1} but values up to 11 mg · l^{-1} were measured.

Growth

The growth of *F. sultana* takes place by budding through the whole year. In fall 1–2 cm thick colony mats covered the underside of the stones. During October and November these mats underwent destruction in their oldest parts and gradually drifted off from the substrate. The drift of colony fragments resulted after some weeks in almost clean stone surfaces regarding colonies, but scattered cystids, still alive, were present during the whole winter. Statoblasts produced in cystids attached to the stones were also recorded.

The germination of statoblasts started in the middle of March at still low water temperature (<4 °C). The development was very slow in the beginning, but during April the temperature raised to about 8–12 °C, resulting in an increased growth of *F. sultana.* A great number of small colonies were now developing both from statoblasts and overwintered individuals ready to utilize the spring

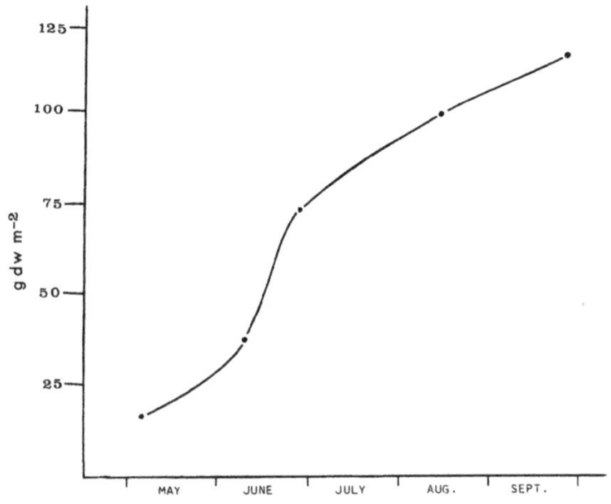

Fig. 1. Accumulation of biomass, as dry weight, of *F. sultana* during the growing season.

Table 1. Mean, maximum and minimum values of chemical parameters in the water from Lake Ulsetstemma.

	Mean	Max.	Min.
pH	5.53	6.1	5.0
H_{20} $\mu S \cdot sm^{-1}$	42.3	58	25
Ca mg · l^{-1}	1.6	2.3	0.9
Mg mg · l^{-1}	0.53	0.7	0.4
SO_4 mg · l^{-1}	4.07	4.5	3.53
PO_4 $\mu g \cdot l^{-1}$	<2	5	<2
NO_3-N $\mu g \cdot l^{-1}$	40	95	<10
Al $\mu g \cdot l^{-1}$	50	80	30
Partic. org. m. (POM) mg · l^{-1}	7	11	<5
Dissolved org. m. (DOM) mg · l^{-1}	<5	<5	<5
Turb. F.T.U.	2.03	5.5	0.52
Colour mg Pt · l^{-1}	69	143	25

bloom of algae in May. During May the colonies occupied most of the free underside of the stones and in June colony branches projected out from the substrate.

From now on the colonies showed the highest rate of accumulation of colony matter due to the continuous budding of new individuals under suitable temperature and nutritional conditions (Fig. 1).

As no invertebrates or vertebrates were observed predating on *F. sultana* in Ulsetstemma the accumulation of biomass can be regarded as net production until death and destruction of the oldest individuals takes place from the end of July.

As shown in Fig. 1 conspicuous amount of colony mass is produced and accumulated during the growth season. The dry weight reaches the value of 117 g · m^{-2} which will be about 1–1.5 kg wet biomass. Little is known about production rates of bryozoans on suitable localities. We know that hundreds of tonnes of them had to be removed from the water pipelines of the big towns in Europe in the beginning of this century (Harmer 1913), but descriptions of production rates are rare. An example is the growth of *Plumatella fungosa* which within a week developed more than 1 kg wet colony mass m^{-2} on gillnets in the eutrophic Lake Esrom (Jonasson 1963). Compaired with the measurements on *F. sultana* in this study the value of *P. fungosa* was very high.

Table 2. Composition in percent of phytoplankton species in water near *F. sultana* colonies. Number in parentheses is the percent value after subtraction of diatoms. + = presence of the species in low number.

Species	May 8	June 25
Cyanophyceae		
Oscillatoria sp. 1 (D = 2 μm)	2.5 (2.7)	+
Oscillatoria sp. 2 (D = 5 μm)	0.3 (0.4)	+
Cryptophyceae		
Cryptomonas erosa	–	6.6 (6.7)
cf. *Rhodomonas minuta*	–	2.6 (2.6)
Dinophyceae		
Unidentified dinoflagellates	0.1 (0.1)	0.7 (0.7)
Chrysophyceae		
Dinobryon divergens	–	1.0 (1.0)
Bacillariophyceae		
Eunotia lunaris	0.4	+
E. praerupta	0.6	0.7
E. veneris	0.2	+
Cymbella spp.	0.1	+
Navicula rhychocephala	0.3	0.1
Navicula spp.	0.5	0.1
Nitzschia spp.	1.2	+
Pinnularia spp.	0.3	+
Tabellaria fenestrata	–	+
T. flocculosa	3.4	0.1
Unidentified diatoms	0.8	+
Euglenophyceae		
Euglena sp.	1.5 (1.6)	+
Chlorophyceae		
Ankistrodesmus sp.	–	+
Chlamydomonas sp.	0.1 (0.1)	45.0 (45.5)
Closterium sp.	+	+
Cosmarium sp.	–	+
Mougeotia sp. (D = 9–14 μm)	0.9 (0.9)	0.4 (0.4)
Scenedesmus sp.	0.1 (0.1)	+
Spirogyra sp. (D = 30 μm)	–	0.4 (0.4)
Staurastrum sp.	–	+
Unidentified flagellates/monads		
Flagellates/monads, <5 μm	47.6 (51.6)	20.1 (20.3)
Flagellates/monads, 5–10 μm	38.1 (41.2)	21.6 (21.8)
Flagellates/monads, 10–30 μm	1.2 (1.4)	0.4 (0.4)

Feeding of F. sultana

Composition of phytoplankton species in water samples and diatoms from sediment samples near the colonies of *F. sultana* are given in Table 2 and 3, respectively.

Algal species found in fecal pellets are listed in Tables 4 and 5.

The digestion efficiency of moss animals is very low and many species of algae are not digested at all (Marcus 1925, 1926; Rüsche 1938). This seemed also to be the case in this study as most of the algae were easy to identify after passing the digestive system.

The composition of algae in the water is generally reflected in the feces of *F. sultana*. In early May flagellates and monads constituted the main food of the animal. No cyanophyceans were found in feces at that time, while *Chlamydomonas* sp. had an increased frequency from 0.1% in water to 2.9% in feces.

High turbidity of the water at that time (5.5–1.2 F.T.U.) corresponded also with high content of particles of detritus minerals in the feces estimated to constitute about 20–30% of the volume of the pellets.

In late June we had a bloom of *Chlamydomonas* sp. (45%) in the water, resulting in 51% in the feces, while the other flagellates/monads at this time constituted about 42 and 34% in water and feces, respectively. At this time we found *Oscillatoria* sp.,

Table 3. Composition in percent of benthic diatoms near the *F. sultana* colonies.

Species	May 8	June 25
Achnathes sp.	1.3	0.7
Cymbella sp.	2.6	2.7
Eunotia lunaris	1.3	3.4
E. polydentula	2.0	1.3
E. praerupta	15.7	30.7
E. veneris	7.3	6.0
Eunotia spp.	–	2.0
Fragilaria undata var. *quadrata*	–	0.3
F. virescens var. *mesolepta*	–	0.3
Fragilaria spp.	12.7	1.0
Frustulia rhomboides	2.0	1.7
Navicula cryptocephala	0.3	4.0
cf. *N. lapidosa*	4.7	4.0
N. rhynchocephala	1.7	4.7
Navicula spp.	1.4	0.7
Nitzschia palea	4.7	7.7
Nitzschia sp.	1.0	3.0
Peronia héribaudi	3.0	4.7
Pinnularia brevicostata	0.3	–
P. interrupta	2.1	1.7
P. subcapitata	–	0.3
Pinnularia spp.	1.3	–
Stauroneis anceps	1.0	1.0
Suriella ovata	1.0	1.7
Tabellaria flocculosa	10.0	11.0
Unidentified diatoms	22.1	5.7

Table 4. Recorded algae minus diatoms in fecal pellets of *F. sultana* given as percent.

Species	May 8	June 5
Cyanophyceae		
Oscillatoria sp. 2 (D = 5 μm)	–	0.7
Dinophyceae		
Unidentified dinoflagellates	–	2.1
Chrysophyceae		
Dinobryon divergens	–	9.3
Chlorophyceae		
Ankistrodesmus sp.	–	1.1
Characium sp.	–	+
Chlamydomonas sp.	2.9	51.3
Cosmarium sp.	0.1	–
Mougeotia sp. (D = 9–14 μm)	–	1.4
Scenedesmus acutus	–	0.2
Staurastrum spp.	0.2	0.2
Xanthidium sp.	0.2	–
Unidentified flagellates/monads		
Flagellates/monads, <5 μm	59.7	6.8
Flagellates/monads, 5–10 μm	34.1	23.1
Flagellates/monads, 10–30 μm	2.9	3.9

Dinobryon divergens, Ankistrodesmus sp., *Mougeotia, Scenedesmus acutus* and *Staurastrum* spp. in the pellets. Of these *D. divergens* had a frequency of 1 and 9% respectively in water and feces.

The amount of benthic diatoms free in the water is low compaired with the normal planktonic species. However, the diatoms are regularly used as food (Table 5), with at least 27 species/groups identified in pellets. Of these, the species *Eunotia praerupta, Nitzschia palea* and *Tabellaria flocculosa* are the most eaten which also dominate the benthos samples (Table 3). Other frequent species in pellets are *E. lunaris, E. veneris, Navicula rhynchocephala, Peronia héribaudi* and *Pinnularia* spp.

The composition of benthic diatoms in feces of *F. sultana* seems generally to reflect the relative composition of the different species at the locality and especially how easy the species can come into drift. An example on the opposite is the occurrence of *Fragilaria* spp., which are tightly bounded to the substrate, are never found in feces in spite of a presence of nearly 13% at the locality in early May.

It is suggested that *F. sultana* eat all particles that can pass through the mouth opening. As examples cells of *Mougeotia* sp. with a length up to 244 μm or rotatorians at about the same size have been eaten. It is observed that the tentacles of *F. sultana* fold around and enclose big active particles until they are swallowed.

Long chains of for instance *T. flocculosa* or greater parts of mats consisting of cyanophyceans entering the lophophore will usually be thrown out again. In such cases the tentacles bend outwards and the algal mass pass away.

As such nutrition is difficult to handle and swallow, it can if not discriminated by the animal, result in clogging of the whole filtering apparatus.

Particle size

A special study regarding the size of the eaten diatoms was made. The species found in water samples and in feces in May and June were grouped into 5 different size categories. The relative occurrence of these categories in water and feces is shown in Figs. 2 and 3.

Diatoms of the categories 0–20 and 20–40 μm dominated both in May and June in the water and constituted together 75–90% of this size fraction. The composition of diatoms in feces of *F. sultana*

Table 5. Recorded diatoms in pellets of *F. sultana* given as percent.

Species	April 27	May 8	May 25	June 10	June 25
Eunotia arcus	–	–	1.3	0.7	–
E. diodon	–	–	–	0.3	0.3
E. lunaris	4.0	4.3	2.6	3.0	1.0
E. polydentula	3.3	1.0	–	–	0.7
E. praerupta	53.7	29.1	25.5	33.8	36.0
E. veneris	2.3	7.4	4.6	7.0	8.3
Cymbella sp. 1	–	0.3	1.0	0.7	–
Cymbella sp. 2	0.7	1.7	0.3	0.7	0.3
Frustulia rhomboides	2.3	0.6	1.3	–	0.7
Navicula cryptocephala	0.7	0.7	–	0.3	0.7
cf. *N. lapidosa*	–	–	–	4.0	1.3
N. pupula var. *elliptica*	–	–	–	1.0	–
N. rhynchocephala	1.7	1.7	0.3	4.3	1.7
Navicula spp.	1.0	1.0	7.0	0.7	–
Neidium bisulcatum	0.3	–	–	–	–
Neidium sp.	0.3	–	–	–	0.3
Nitzschia obtusa var. *scalpelliformis*	–	1.3	0.7	0.7	–
N. palea	2.7	5.0	11.9	17.6	19.0
Nitzschia sp.	–	–	0.3	–	0.3
Peronia héribaudi	3.3	4.0	0.7	1.3	1.3
Pinnularia interrupta	–	4.7	1.0	3.3	3.3
P. subcapitata	1.3	2.3	1.3	5.0	3.7
P. viridis var. *sudetica*	–	–	–	0.3	–
Pinnularia spp.	–	0.7	0.3	0.3	–
Suriella ovata	0.3	0.3	–	1.3	0.3
Tabellaria fenestrata	1.0	1.3	1.0	1.0	1.0
T. flocculosa	17.3	24.7	30.5	6.0	15.0
Unidentified diatoms	2.7	7.0	5.3	7.0	4.7

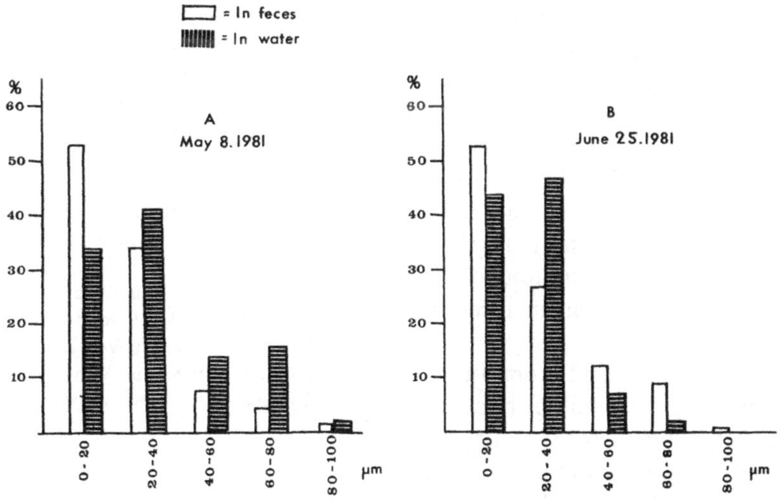

Fig. 2. Frequency in percent of five length groups of diatoms found in feces of *F. sultana* and in water samples in May (A) and June (B).

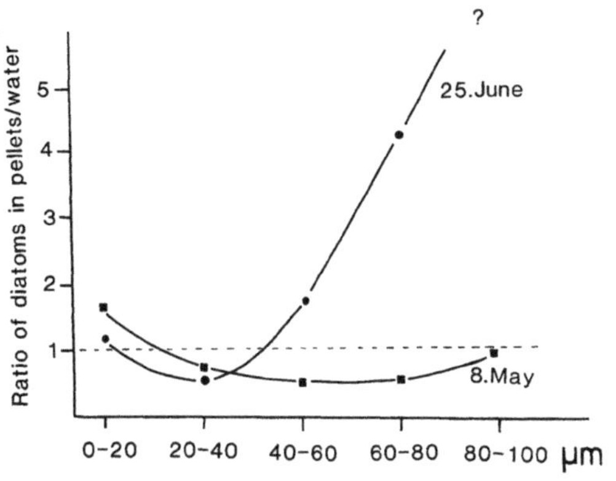

Fig. 3. The ratio between the relative number of diatoms in feces of *F. sultana* on the relative number in water for five different size categories. Broken line indicate the 1:1 ratio.

show a similar trend both in May and June, but variations in filtering different particle sizes were recorded.

In May the feces contained an elevated number of diatoms <20 μm and a reduced number of the larger, compaired with the composition in water (Figs. 2A and 3). In June the same trend was found regarding diatoms in the categories 0–20 and 20–40 μm, but the larger now showed an increase frequency in feces compaired to the presence in water (Figs. 2B and 3). It seemed clear that young and old *F. sultana* individuals have different feeding strategies.

It was suggested that the wave action of the cilia on the tentacles and elsewhere on the lophophore were responsible for the filtration of the smaller particles (<20 μm). It was also observed that bigger particles (100–200 μm) were caught by the tentacles themselves either by enclosing the particles or by striking/pushing them down to the mouth.

As the number and length of the tentacles increase with the age of the individuals (17 → 27), the older individuals may be more clever and have more 'experience' in handling bigger particles than the younger. This may be an explanation to the relative increase of diatoms >40 μm in feces in June.

The reduced efficiency in catching diatoms in the size between 20–40 μm may on the other hand be due to these particles being too big for the cilia and too small for the tentacles. Diatoms of that size may therefore to a greater extent than the others escape between the tentacles.

Conclusion

The study of *F. sultana* has showed that this species can stay in humic water with pH down to 5.0. In absence of predation *F. sultana* can build up great biomass utilizing the drift of nutrition from the water body. During algal blooms and suitable temperature the animals have rapid growth rate feeding on particle sizes from the smallest up to about 240 μm in length. The wide of the particles must however not be greater than the mouth opening.

As *F. sultana* has a very poor digestion the animal has to maintain a high turnover of nutrition particles. Colonies in rapid growth may therefore play an important role regarding breakdown and recycling of nutrition in the freshwater ecosystem as well as structuring the phytoplankton community by the selective feeding observed.

References

Harmer, S. F., 1913. The Polyzoa of waterworks. Proc. zool. Soc. Lond. 31: 246–257.

Heding, S. G., 1938. The ecology of the fresh-water Polyzoa in East Anglia. J. r. Microsc. Soc. Lond. 47: 135–142.

Hustedt, F., 1930. Bacillariophyta: Die Süsswassers-Flora Mitteleuropas. Jena 10: 466 pp.

Jonasson, P. M., 1963. The growth of Plumatella repens and P. fungosa (Bryozoa Ectoproeta) in relation to external factors in Danish eutrophic lakes. Oikos 14: 121–137.

Marcus, E., 1925. Bryozoa. In: Schulze, P. (ed.). Biologie der Tiere Deutschlands. Liefg. 14, Teil 47, 1–46 pp.

Marcus, E., 1926. Beobachtungen und Versuche an lebenden Süsswasserbryozoen. Zool. J. 52. 279–350 pp.

Marcus, E., 1940. Mosdyr (Bryozoa eller Polyzoa) Danmarks Fauna 46, 401 pp.

Prescott, G. W., 1970. Algae of the Western Great Lakes Area. Rev. Ed. Wm. C. Brown, Dubuque, Iowa, 977 pp.

Raddum, G. G., 1973. Mosdyr-faunaen (Bryozoa) i diverse ferskvann på Vestlandet, biologi og taxonomi. Thesis, Univ. Bergen, 153 pp.

Rüsche, E., 1938. Nahrungsaufnahme und Nahrungsauswertung bei Plumatella fungosa (Pallas). Arch. Hydrobiol. 33: 271–293.

Utermöhl, H., 1931. Neue Wege in der quantitativen Erfassung des Planktons. (Mit besonderer Berücksichtigung des Ultraplanktons.) Verh. int. Ver. Limnol. 5: 567–596.

Utermöhl, H., 1958. Zur Vervollkommung der quantitativen Phytoplankton-methodik. Mitt. int. Ver. Limnol. 9: 1–38.

The food preference of vendace *(Coregonus albula)* in South Swedish forest lakes including the predation effect on zooplankton populations

Stellan F. Hamrin
Institute of Limnology, University of Lund, Box 3060, S-220 03 Lund, Sweden

Keywords: forest water ecosystems, *Coregonus albula,* food selection, predation effects, *Bosmina coregoni*

Abstract

Vendace (*Coregonus albula,* L.) is in the southern part of its distribution area during the summer period restricted to hypolimnion. Food is dominated by *Bosmina coregoni* and *Daphnia* sp. The selectivity index is highest for large cladocerans and large copepods.

The pelagic crustacean fauna in vendace lakes is dominated by small species like *Bosmina coregoni, Daphnia cristata, Thermocyclops* and *Mesocyclops* sp. and *Eudiaptomus* sp. During periods of maximum vendace abundance only *Bosmina coregoni* is left of the cladocerans, while the abundance of cyclopoid copepods increases.

Introduction

Size-dependent predation by planktivorous fish has a strong influence on species composition of the zooplankton fauna. Numerous investigations have shown how increase in abundance of planktivorous fish favours abundance of small Cladocera and Copepoda and how presence of large zooplankton species is dependent upon the absence of such fish (Hrbácek 1962; Brooks & Dodson 1965; Galbraith 1967; Brooks 1968; Stenson 1972; Nilsson & Pejler 1973; Andersson *et al.* 1978). Less attention, however, has been paid to the quantitative effects of short-term changes in the size of planktivorous fish populations (De Bernardi & Giusanni 1975; Gophen & Landau 1977). The present investigation describes such a case with vendace *(Coregonus albula* L.)* being the totally dominating planktivore.

Lake description and methods

Lake Bolmen (area 181 km^2) is an oligotrophic lake (transparency 4–6 m, specific conductivity 60–70 μS $\cdot$ cm^{-1}, pH 6,4–7,0 and total phosphorous 6–16 μg $\cdot$ l^{-1}) situated in the southwestern part of Sweden. The lake has three basins of which the biggest used for this study has a maximum depth of 37 m. Thermal stratification takes place at depths between 10 and 20 m during summer. The fish fauna is dominated by vendace and perch (*Perca fluviatilis* L.). Other important species are pikeperch (*Stizostedion lucioperca* L.), pike (*Esox lucius* L.),

* Svärdson (1979) has for 'international discussion' proposed the names ciscoe or Baltic ciscoe as the English vernacular equivalents for the species *Coregonus albula.* Svärdson also suggests that *Coregonus albula* populations in Great Britain with the vernacular name vendace might be a true species, which however is closely related to *Coregonus albula.* As the American *Coregonus artedii* definitely is a species of its own – and less related to *Coregonus albula* than is the case for *Coregonus vandesius* – it seems most reasonable to give the Scandinavian *Coregonus albula* populations the English vernacular name vendace instead of the vernacular name ciscoe originally used for *Coregonus artedii.* If there is a need to use different vernacular names for *Coregonus albula* and *Coregonus vandesius* the former could be called Baltic vendace. In this article however the English vernacular name vendace refers to the species *Coregonus albula.*

Hydrobiologia 101, 121–128 (1983).

whitefishes (*Coregonus wartmanni* Malm and *Coregonus nilssoni* Val., sensu Svärdson 1979) and roach (*Rutilus rutilus* L.).

Of the other lakes studied for comparison purposes Lake Ivösjön (area 54 km², max. depth 50 m) is oligohumic and mesotrophic, Lake Skärlen (area 3 km², max. depth 27 m) is oligohumic and oligotrophic and Lake Gyltigesjön (area 0.4 km², max. depth 18 m) is polyhumic and oligotrophic. All lakes are thermally stratified during summer with well-developed hypolimnions (for more detailed information see Hamrin 1979).

Vendace was caught in October each year at a temperature between 10 and 11 °C and in July at a hypolimnion temperature between 10 and 13 °C. Monofilament gill nets were used with a mesh size (knot to knot) of 10 and 15 mm. Gill nets were set at 9:00 at three depths (2-5, 11-14, 21-24 m) above the deepest part of the lake and replaced 24 hours later. The catch was weighted with the share of the total volume of the lake represented by the fishing depth. Correction for length selectivity of the gill nets has been made (Hamrin 1979).

Zooplankton samples were taken during daytime by a 5 l metal sampler (Rodhe sampler) at 0, 2, 6, 12 and 25 m depth in July and October. The samples from different depths have been weighted according to the method used for vendace.

Results

Zooplankton

The cladocerans of Lake Bolmen were dominated by *Bosmina coregoni* (Table 1). Other abundant species were *Daphnia cristata* and *Holopedium gibberum*. The abundance of *Leptodora kindti* and *Bythotrephes cederströmi* was very low. Among copepods *Mesocyclops leuckarti, Eudiaptomus gracilis* and *Thermocyclops oithonoides* were dominating. Large species like *Eurytemora lacustris* and *Heterocope appendiculata* were rare.

Bosmina coregoni had a maximum in early summer (Fig. 5), while copepods were dominating during late summer.

Vendace

Vendace is the most common fish species in Lake Bolmen (Hamrin 1979). The population, which is strongly dominated by age class 0+, exhibits cyclic population oscillations with a strong year class every second year (Table 2).

During summer stratification all age classes of vendace are restricted to the hypolimnion (Fig. 1). Only, exceptionally, during cold and windy summers with weak stratification (e.g. 1974), vendace can be found in surface waters. Echosounding studies have shown, that age class 0+ is somewhat closer to the thermocline than older fish and might penetrate the epilimnion during diurnal vertical migrations (Hamrin 1979).

The gut content of vendace was dominated by *Bosmina coregoni, Holopedium gibberum* and *Daphnia* sp. (Table 3). The October differences in the gut contents of Lake Bolmen vendace were small (Fig. 2). In July, however, a correlation between the gut content and the strength of stratification could be found. In 1970, when the difference between the epilimnion and hypolimnion tempera-

Table 1. Abundance of planktonic crustaceans in Lake Bolmen 1970–1975. The values (ind. · l^{-1}) represent the mean of 5 samples (5 liter each) between 0 and 25 m depth in July and October. Of the species with abundances less than 0.1 ind. · l^{-1} only those important to vendace have been given.

Cladocera	ind. · l^{-1}	Copepoda	ind. · l^{-1}
Bosmina coregoni	4.7	*Mesocyclops leuckarti*	2.9
Daphnia cristata	0.9	*Eudiaptomus gracilis*	2.2
Holopedium gibberum	0.6	*Thermocyclops oithonoides*	1.6
Daphnia galeata	0.3	*Eurytemora lacustris*	0.2
Chydorus sphaericus	0.2	*Cyclops lacustris*	0.1
Diaphanosoma brachyurum	0.2	*Heterocope appendiculata*	0.0
Daphnia longispina	0.1		
Bosmina longirostris	0.1		
Ceriodaphnia quadrangula	0.1		
Limnosida frontosa	0.1		
Leptodora kindtii	0.0		
Bythotrephes cederströmi	0.0		

Table 2. Catch per unit effort (ind.-%) of vendace (year classes 0+ and older) in Lake Bolmen in October 1970–1975.

	Year					
	1970	1971	1972	1973	1974	1975
Catch/unit effort						
Age class 0+	2.6	18.6	6.6	100.0	10.7	78.6
Age class ≥1+	7.9	0.6	4.1	0.2	1.6	0.6

ture was only 2 °C, *Bosmina coregoni* was an important food item, while during summers with more pronounced stratification (1972, 1973, 1975; temperature differences 6–8 °C) *Bosmina coregoni* was almost missing in the guts.

The Ivlev electivity index (Table 4) was highest for the large cladocerans *Holopedium gibberum, Bythotrephes cederströmi* and *Leptodora kindti.* The index for *Bosmina coregoni* was also positive, while the indices for copepods were negative.

Relation between abundance of vendace and planktonic crustaceans

During the years 1970–1975 the abundances of the dominating planktonic crustaceans were negatively correlated with the relative abundance of vendace (Fig. 3). The abundance of *Daphnia cristata* ($p < 0.01$), *Bosmina coregoni* ($p < 0.05$) and *Holopedium gibberum* ($p < 0.05$) showed a negative correlation with the relative abundance of vendace, while the correlation with the abundance of *Eudiaptomus gracilis* was negative but not statistically significant ($p < 0.1$). The abundance of the

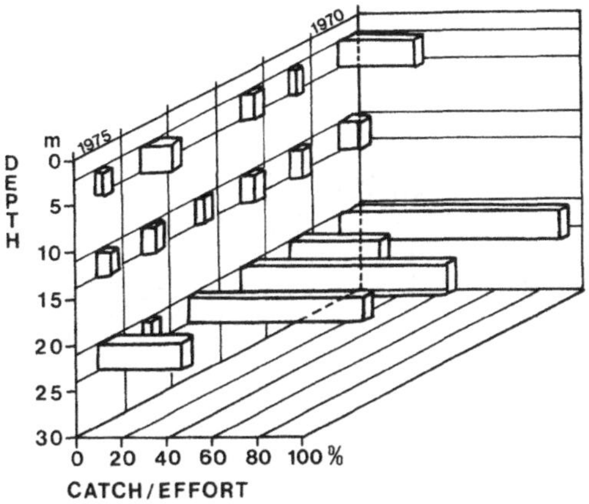

Fig. 1. Catch per effort corrected for gill net selectivity (Hamrin 1979) in Lake Bolmen in July 1970–1975. Each bar represents the mean of 2 (1970, 1971) or 3 24-hour fishing periods with 270 m² gill net each. 100% equals 230 fish.

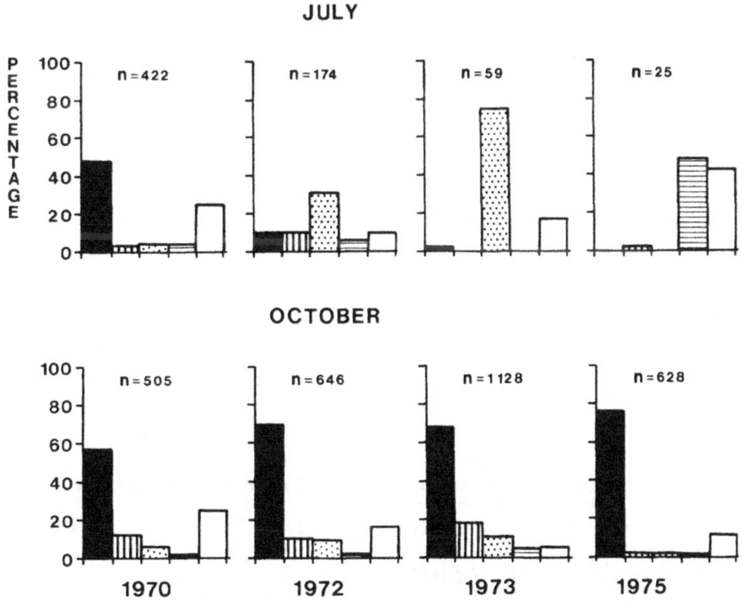

Fig. 2. Relative food composition of vendace in Lake Bolmen in July (11–24 m depth) and October (2–4 m depth) 1970 (98 and 20 sampled stomachs resp.), 1972 (45 and 38 resp.), 1973 (30 and 20 resp.), 1975 (53 and 20 resp.). The years represent a series of increasing differences between epilimnion and hypolimnion temperature in July. *n* indicates the mean number of food objects per fish. The five different bars represent *Bosmina coregoni* (black), *Daphnia cristata, Holopedium gibberum, Bythotrephes cederströmi* and copepods (open) resp.

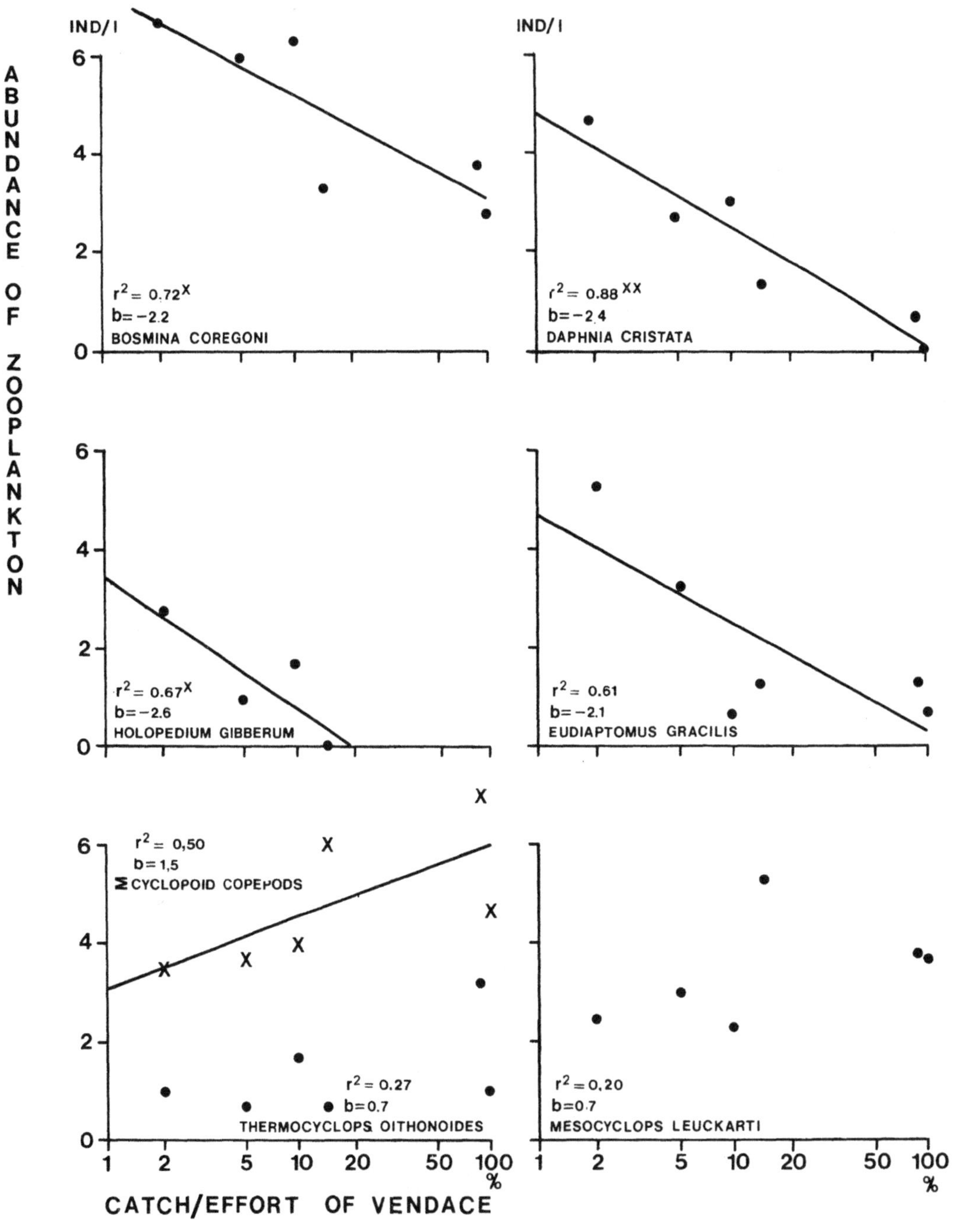

Fig. 3. Relation between catch/effort of yearling vendace and abundance of dominating planktonic crustaceans in Lake Bolmen 1970–1975. The zooplankton abundance is based on 5 5-l samples between 0 and 25 m depth taken in July and October. Abundance of vendace is based on 2–3 24-hour fishing periods (270 m² gill net distributed over 3 depths) in October. x and xx indicate significant correlation coefficients on the 95 and 99% level, resp.

Table 3. Composition (percentage) of food of vendace (150–250 mm) in lakes northern Bolmen (N.B., 1970–1975), southern Bolmen (S.B., 1970–1975), Skärlen (Sk, 1972), Gyltigesjön (Gy, 1972, 1975) ànd Ivösjön (Iv, 1976). The sampled depths were those preferred by vendace.

Month	July					October				Mean
Lake	N.B	S.B	Sk	Gy	Iv	N.B	S.B	Sk	Iv	
Depth (m)	8-11	11-24	15-25	13-18	10-30	2-5	2-5	2-5	2-5	
Nos. sampled	108	160	8	16	65	49	49	10	54	
Nos. of food objects	260	90	4 400	2 700	951	850	450	4 600	121	
Bosmina coregoni	38	23	20	–	48	70	69	35	7	34
Holopedium gibberum	11	30	80	–	–	4	9	35	–	19
Daphnia cristata	8	5	–	–	1	15	3	30	–	7
Daphnia longispina	–	–	–	51	+	–	–	–	3	6
Bythotrephes cederströmi	3	13	–	–	–	3	6	–	–	3
Limnosida frontosa	13	3	–	–	+	2	–	–	–	2
Chydorus sphaericus	–	–	–	–	+	3	2	–	–	1
Leptodora kindti	1	–	–	–	+	–	2	–	–	0
Calanoid copepods	5	9	–	4	15	4	7	–	83	14
Cyclopoid copepods	3	3	–	44	21	4	2	–	4	9

cyclopoid copepods (almost exclusively *Mesocyclops leuckarti* and *Thermocyclops oithonoides*) was high during years with high abundance of vendace, but the correlation was not significant ($p < 0.1$).

The correlation between the total abundance of all crustacean species and the relative abundance of vendace was higher than for the individual species (Fig. 4). During years with high abundance of vendace the cyclopoid copepods were dominant in the plankton community. Of other species only *Bosmina coregoni* was abundant. During years with low abundance of vendace cladocerans dominated over copepods.

Discussion

The dominating food objects of vendace was *Bosmina coregoni* and different species of daphnids (in oligotrophic lakes *D. cristata*). This seems to be true for all parts of the distribution area of vendace (Hakkari 1978; Hamrin 1979) and also for other planktivores like *Osmerus eperlanus* and *Alosa* sp. (Hutchinson 1971; Mistautaite 1974; Hakkari 1978). This preference of planktivores for cladocerans is probably due to the low evasion ability of most cladocerans (O'Brien 1979).

The electivity index was highest for large cladocerans and large copepods because of the greater

Table 4. Electivity index (Ivlev) of vendace with 95% confidence interval in Lake Bolmen, Lake Pluszne (Marciak 1962), Lake Konnevesi (Hakkari 1972) and Lake Mälaren (Vallin 1969). n = total number of indices used for computation of mean and confidence interval. The abundance of species missing in the plankton sample has been put to 0.1%. Only those indices have been included, which are based on abundances of food organisms exceeding 10% in either the water sample or the food sample.

Fish size	100-130 mm	150-250 mm					
Lakes	Bolmen	Bolmen	Pluszne	Konnevesi	Mälaren	Total	n
Bosmina coregoni	+0.19 ± 0.23	+0.06 ± 0.27	+0.36 ± 0.52	+0.22 ± 0.92	+0.82 ± 0.28	+0.27 ± 0.18	33
Daphnia cristata	-0.37 ± 0.92	0.57 ± 0.18	+0.04 ± 0.30	-0.47 ± 0.64	+0.05 ± 1.16	-0.31 ± 0.15	34
Holopedium gibberum	+0.35	+0.46 ± 0.08				+0.46 ± 0.08	15
Bythotrephes cederströmi	-0.15	+0.55 ± 0.21				+0.55 ± 0.21	17
Leptodora kindti			+0.99 ± 0.02	+0.96 ± 0.07		+0.98 ± 0.02	7
Eudiaptomus gracilis } *Eudiaptomus graciloides*		-0.31 ± 0.44	-0.62 ± 0.49	-0.51		-0.41 ± 0.30	16
Thermocyclops oithonoides } *Mesocyclops leuckarti*	-0.48 ± 0.39	-0.67 ± 0.18	-0.25 ± 0.42			-0.51 - 0.18	24
Limnocalanus macrurus				+0.02 ± 0.60	-0.09 ± 0.65	-0.04 ± 0.40	14

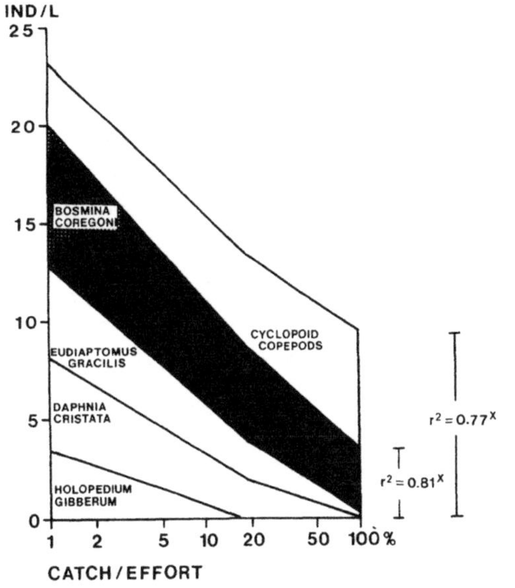

Fig. 4. The relation in Lake Bolmen between relative abundance of yearling vendace and the abundance of the dominating pelagic crustacean species based on the results in Fig. 3. x indicates significant correlation coefficient on the 95% level.

reactive distances for large preys (cf O'Brien 1979; Eggers 1977; Gardner 1981). Predation from vendace is therefore a likely reason for the very strong reduction of *Daphnia longispina* and *Heterocope appendiculata* in vendace lakes. The abundances of the large cladocerans *Bythotrephes cederströmi* and *Leptodora kindti* are however not dependent upon presence of vendace (Ekström 1975). This is supported by the fact that these cladocerans are rare in vendace guts but a dominant food item for roach and whitefish caught in the same area (Nilsson 1974; Hamrin 1979).

The dominance of *Bosmina coregoni, Daphnia cristata, Holopedium gibberum* and *Eudiaptomus gracilis* in the food of vendace strongly indicates, that the inverse relationship found between the abundance of vendace and these crustaceans was caused by predation from vendace. According to the size-selectivity and size-efficiency hypothesis (Hall *et al.* 1976) these species should in lakes with a high predation pressure be replaced by smaller filtrators. An increase in abundance of rotifers has been found in enclosures with high abundance of planktivores (Andersson *et al.* 1978). Rotifers are, however, too small to be consumed by vendace. The decline in population of crustacean filtrators was accompanied by an increase in the population of cyclopoid copepods. Hrbácek (1962) and Warshaw (1972) have earlier found correlations between fish and cyclopoid copepod abundances. This correlation could be an effect of higher abundance of rotifers – important food items for these copepods (Nauwerck 1963) – in the absence of cladocerans, thus being an example of complementary feeding niches sustained by size-selective predation (Dodson 1970).

During periods of high vendace abundance *Daphnia cristata* and *Holopedium gibberum* were almost eliminated and *Eudiaptomus gracilis* was reduced to concentrations below 0.5 ind. l^{-1}. Of the cladocerans only *Bosmina coregoni* was left. Its abundance was lowered but its relative share of the crustacean community was strengthened. *Bosmina coregoni* has been found to be the dominating cladoceran in lakes with high abundance of planktivores like *Alosa* sp. (Warshaw 1972) and cyprinids (Grygierek 1973; Maltsman 1974). As *Bosmina coregoni* is selected by e.g. predators like vendace

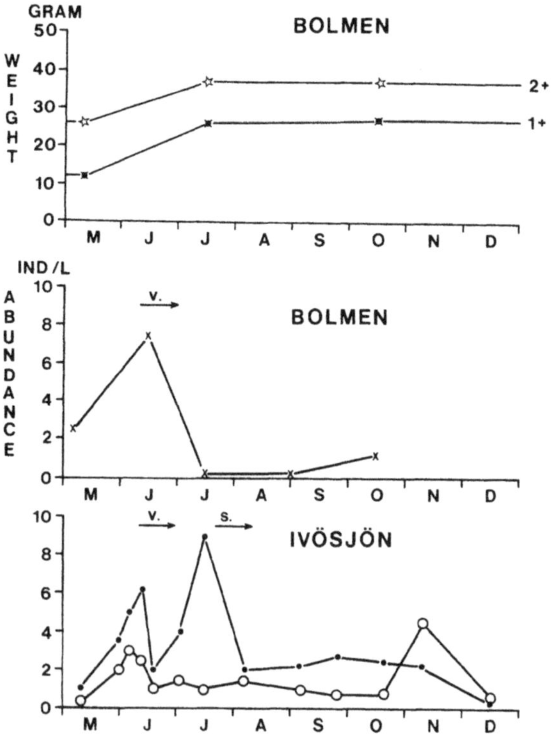

Fig. 5. Weight of vendace in Lake Bolmen (above) and abundance of *Bosmina coregoni* in Lake Bolmen 1977 (0–25 m) and in Lake Ivösjön 1979 (points 0–10 m; circles 30–45 m). v. and s. indicate the approximate pelagic appearance of crustacean-eating yearlings of vendace and smelt, resp.

and smelt, its ability to maintain its position in the planktonic community during high predation pressure is remarkable. *Bosmina coregoni* has obviously found a life strategy much more successful than other cladocerans of similar size. One part of this strategy might be its short generation time (Hakkari 1978; O'Brien 1979), contrasting to the strategy of replacement cycles and different morphs adopted by *Bosmina longirostris* (Kerfoot *et al.* 1980; Nilssen *et al.* 1980).

It must be of very great value to vendace that *Bosmina coregoni*, one of its most preferred food items, is the cladoceran species most resistant to predation. *Bosmina coregoni* has a very rapid population increase in early summer (Fig. 5). The period of peak abundance coincides with the period of maximum growth for all age classes of vendace. It is only during this period that the growth of vendace is density-independent (Hamrin 1979). It is also during this period that the vendace fry becomes dependent upon crustaceans as a food source.

The predation from vendace could be one reason for the rapid decline of the *Bosmina* population during summer. During the summer stratification period the vendace population is concentrated in the hypolimnion of the lake. The fry and occasionally also the adult vendace perform vertical, diurnal migration to the epilimnion (Hamrin 1979). The predation pressure in this area must therefore be very intense.

In contrast to the situation in Lake Bolmen the populations of *Bosmina coregoni* in Lake Ivösjön (Fig. 5) and in Lake Erken (Nauwerck 1963) have two peaks in summer followed by a rapid decline. The two last-mentioned lakes have strong populations of smelt, which fry develop much later than the fry of vendace. The period of the first decline of the *Bosmina* population in these lakes coincides with the pelagic appearance of vendace and perch fry and the second period of decline with the appearance of smelt fry. It is also notable that the abundance of *Bosmina coregoni* during summer is lower in hypolimnion – where the predation pressure is high – than in epilimnion, and that the bottom water abundance of *Bosmina coregoni* rizes in November, when the predators leave that biotop for the surface waters. Keast (1980) noticed a decline in early summer in the population of *Bosmina longirostris*, that was correlated with heavy predation from juvenile fish (mainly Percidae).

The vendace predation on planktonic crustaceans in the area connected with the hypolimnion will probably create a gradient in abundance of crustaceans species towards the littoral zone. The abundance will be lowest in and above the hypolimnion and highest in sheltered, shallow bays out of reach for the vendace during summer. It has been suggested that this type of different predation pressures within a lake is the reason for the development of different phenotypes in *Bosmina longirostris* (Kerfoot & Peterson 1980; Nilssen *et al.* 1980).

Acknowledgements

Thanks are due to Dr Bruno Berzins for zooplankton counts, to Gunnar Andersson and Lennart Persson for helpful comments, and to Bob Petersen for correcting the language.

References

Andersson, G., Berggren, H., Cronberg, G. & Gelin, C., 1978. Effects of planktivorous and benthivorous fish on organisms and water chemistry in eutrophic lakes. Hydrobiologia 59: 9–15.

Brooks, J. L., 1968. The effects of prey size selection by lake planktivores. Syst. Zool. 17: 273–291.

Brooks, J. L. & Dodson, S. I., 1965. Predation, body size, and composition of plankton. Science 150: 28–35.

De Bernardi, R. & Giusanni, G., 1975. Population dynamics of three cladocerans of Lago Maggiore related to predation pressure by a planktophagous fish. Verh. int. Ver. Limnol. 19: 2901–2912.

Dodson, S. I., 1970. Complementary feeding niches sustained by size-selective predation. Limnol. Oceanogr. 15: 131–137.

Eggers, D. M., 1977. The nature of prey selection by planktivorous fish. Ecology 58: 46–59.

Ekström, Ch., 1975. Djurplankton. Rödingsjöar söder om Dalälven. Inform. Inst. freshwat. Res. Drottningholm 7: 49–53.

Galbraith, M. G., Jr., 1967. Size-selective predation on Daphnia by rainbow trout. Trans. amer. Fish. Soc. 96: 1–10.

Gardner, M. B., 1981. Mechanisms of size selectivity by planktivorous fish: a test of hypothesis. Ecology 62: 571–578.

Gophen, M. & Landau, R., 1977. Trophic interactions between zooplankton and sardine Mirogrex terraesanctae populations in Lake Kinneret, Israel. OIKOS 29: 166–174.

Grygierek, E., 1973. Zooplankton w przesadkach I. Rocz. nauk Roln. 95 H 3: 7–33.

Hakkari, L., 1978. On the productivity and ecology of zooplankton and its role as food for fish in some lakes in central Finland. Biol. Res. Dep. Univ. Jyväskylä 4: 1–87.

Hall, D. J., Threlkeld, S. T., Burns, C. W. & Crowley, P. H., 1976. The size-efficiency hypothesis and the size structure of zooplankton communities. Annu. Rev. Ecol. Syst. 7: 177–208.

Hamrin, S. F., 1979. Population dynamics, vertical distribution and food selection of vendace, Coregonus albula, in south Swedish lakes. Thesis, Univ. Lund, 195 pp.

Hrbáček, J., 1962. Species composition and the amount of zooplankton in relation to the fish stock. Rozpr. čsl. Akad. Věd. Řada 72 10: 1–80.

Hutchinson, B. P., 1971. The effect of fish predation on the zooplankton of ten Adirondack lakes, with particular reference to the alewife, Alosa pseudoharengus. Trans. am. Fish. Soc. 100?: 325–335.

Keast, A., 1980. Food and feeding relationships of young fish in the first weeks after the beginning of exogenous feeding in Lake Opinicon, Ontario, Envir. Biol. Fish 5: 305–314.

Kerfoot, W. Ch. & Peterson, C., 1980. Predatory copepods and Bosmina: Replacement cycles and further influences of predation upon prey reproduction. Ecology 61: 417–431.

Maltsman, T. S., 1974. Zooplankton development in heavily stocked fishponds. J. Hydrobiol. 10: 26–30.

Marciak, Z., 1962. Seasonal variations in feeding habits and growth of Coregonus albula in Pluszne Lake. Roczn. Nauk roln. B. 81: 335–357.

Mistautaite, V. & Kublickas, A., 1975. Morphological and ecological variability of smelt in Lithuanian waters. Liet. TSR Mokslu Akad. Darb. C 3 (English summary).

Nauwerck, A., 1963. Die Beziehungen zwischen Zooplankton und Phytoplankton im See Erken. Symb. bot. ups. 17, 163 pp.

Nilssen, J. P., Halvorsen, G. & Melåen, J. G., 1980. Seasonal divergence of Bosmina morphs. Int. Revue ges. Hydrobiol. 65: 507–516.

Nilsson, N.-A., 1974. Fiskens näringsval i öppna Vänern. Inform. Inst. freshwat. Res. Drottningholm. 17: 1–57.

Nilsson, N.-A. & Pejler, B., 1973. On the relation between fish fauna and zooplankton composition in North Swedish lakes. Rep. Inst. freshwat. Res. Drottningholm 53: 51–77.

O'Brien, W. J., 1979. The predatory-prey interaction of planktivorous fish and zooplankton. Am. Scient. 67: 572–581.

Stenson, J. A. E., 1972. Fish predation effects on the species composition of the zooplankton community in eight small forest lakes. Rep. Inst. freshwat. Res. Drottningholm 52: 132–148.

Svärdson, G., 1979. Speciation of Scandinavian Coregonus. Rep. Inst. freshwat. Res. Drottningholm 57: 1–95.

Vallin, S., 1969. Siklöjans näringsbiologi i Lambarfjärden, Mälaren. Inform. Inst. freshwat. Res. Drottningholm 7: 1–57.

Warshaw, S. J., 1972. Effects of alewifes (Alosa pseudoharengus) on the zooplankton of Lake Wononskopomuc, Connecticut. Limnol. Oceanogr. 17: 816–825.

Food and food selection of cisco (*Coregonus albula* L.) in a dysoligotrophic lake

M. Viljanen
Karelian Institute, Section of Ecology, University of Joensuu, Box 111, 80101 Joensuu 10, Finland

Keywords: forest water ecosystems, cisco, food, food selection, zooplankton, distribution

Abstract

With minor exceptions, cisco *(C. albula)* in Lake Suomunjärvi fed on cladocerans and copepods. The food habits were different among two size classes of fish. The composition of cisco diet changed at different depths and times of the day, but the distribution of zooplankton was usually very similar in each period irrespective of the depth or diel period. *Daphnia, Bosmina, Cyclops scutifer* and *Heterocope appendiculata* were the main diet of cisco. Calculations of fish electivity indices showed that cisco selected usually large species of cladocerans and copepods and a small cladoceran, *Bosmina coregoni.* Body-size selection was clear for *Daphnia* and *Bosmina.*

Introduction

In spite of the current interest in cisco (*Coregonus albula* L.) management in freshwaters, relatively little attention has been given to studying their food habits, prey selection and competition for food with other fish species.

The food habits of cisco in Finland have been studied recently by Airaksinen (1967), Hakkari (1972, 1978), Viljanen (1972, 1974), Piipponen (1977), and in other countries Patalas (1954), Marciak (1962), Hammar (1968), Vallin (1969), Szypula (1970), Stenson (1972), Hamrin (1979), and Nilsson (1979). The food habits of cisco (*Coregonus artedii* L.) in North America have been studied by Smith (1956) and Engel (1976). According to these papers cisco are primarily planktivorous throughout life feeding usually on small, seasonally abundant zooplankton. The diet consists of mainly cladocerans and copepods. Rotifers are rarely eaten by cisco and are important only for the fry. However, in lakes where large zooplankters *(Limnocalanus, Bythotrephes, Leptodora, Megacyclops, Mysis, Chaoborus)* are abundant they can play a very important role in cisco diet. The abundance of aquatic and terrestrial insects in the diet changes seasonally and they are usually important only during spring and early summer.

It is a well-known fact that most pelagic species of fish display some sort of seasonal and diel vertical and horizontal migrations (Northcote 1978). This is well documented with *C. albula* and *C. artedii* as reviewed by Maitland (1968), Dembinski (1971), Northcote & Rundberg (1971), Engel & Magnuson (1976), Hamrin (1979) and Nilsson (1979).

Many planktivorous fish are known to feed selectively on large species of zooplankton. Under conditions of intense predation, the larger species may be eliminated or reduced to insignificance by the planktivore and then be replaced by smaller species. The replacement of larger species by smaller species has been first described by Hrbáček (1962) and Brooks & Dodson (1965) and after them well documented in many lakes and ponds where planktivorous fish are abundant. However, Gliwicz & Prejs (1977) have shown that factors other than fish predation are responsible for the seasonal and spatial changes of crustacean populations.

Hydrobiologia 101, 129–138 (1983).

Many workers have found Ivlev's electivity index useful in comparing fish food habits with the availability of potential food sources in natural habitats. The electivity index has the advantage of offering an unbiased comparison of percent composition of stomach contents with plankton available, irrespective of the magnitude of the paired values. However, in recent years Jacobs (1974) and Paloheimo (1979) have criticized Ivlev's electivity index as measures of preference by predators of different prey or food types.

The purpose of this investigation was to examine the food habits of cisco *(C. albula)* in Lake Suomunjärvi in 1976. Special attention has been paid to food comparison in different seasons, diel periods, depth zones, size of the fish and the selection of food organisms. Preliminary papers on food habits of cisco in relation to zooplankton in Lake Suomunjärvi in 1974 has been presented by Piipponen (1977) and on zooplankton in 1976 by Latja (1978).

Material and methods

Study area

Lake Suomunjärvi is a mesohumic oligotrophic lake in eastern Finland (63°08'N and 30°47'E) with an approximate area of 650 ha and a mean depth of 5.7 and maximum depth of 27 m. It is divided into southern, central and northern basins by islands and sand banks. The lake is normally ice-covered from early November to mid-May. According to experimental gillnetting the main species of fish inhabiting the lake are: the cisco (*Coregonus albula,* biomass 30.1%), the perch (*Perca fluviatilis,* 26.5%), the whitefish (*Coregonus lavaretus,* 14.1%), the roach (*Rutilus rutilus,* 12.2%), the dace (*Leuciscus leuciscus,* 7.5%) and pike (*Esox lucius,* 5.3%). A summary of the main physicochemical features of the lake has been published by Viljanen & Holopainen (1982).

In 1976 the lake was poorly stratified (Fig. 1). The formation of a metalimnion started in the beginning of June and was at 15 to 18 m during June and at 18 to 20 m during July, August and September. The maximum temperatures in the epilimnion (July and August) were 18–19 °C. The stratification was broken at the end of September. The stratification of oxygen was also very poor. During the ice-free time the values varied in epilimnion from 8.2 to 10.5 mg $O_2 \cdot l^{-1}$ and in hypolimnion from 5.5 to 8.5 mg $O_2 \cdot l^{-1}$.

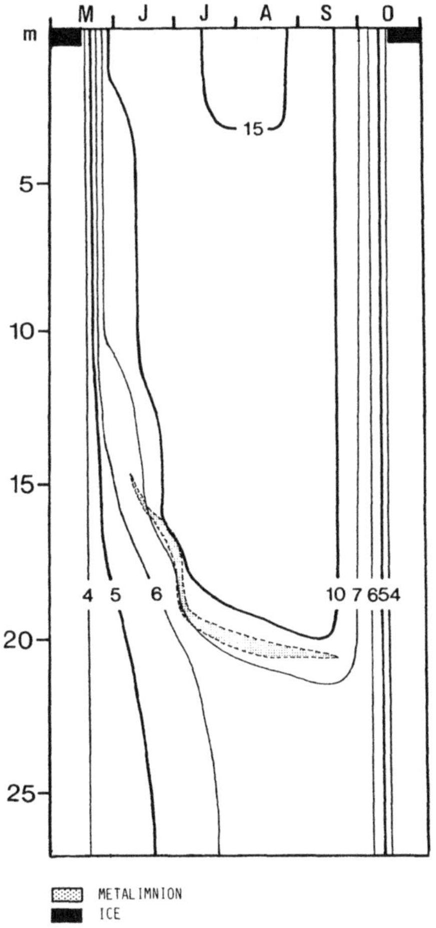

Fig. 1. The central basin, deep station. Temperature of the water in 1976.

Zooplankton

The water samples for zooplankton analyses were taken in the deep station of central basin at every meter in a water column of 0–20 m (from surface to bottom) and pooled into four groups (0–5, 5–10, 10–15 and 15–20 m), which were poured separately through a plankton net with a mesh size of 50 μm. These samples were taken at 6 h intervals through 24 h for June (10/6–11/6 and 23/6–24/6), July (6/7–7/7 and 21/7–22/7), August (4/8–5/8 and 30/8–31/8), September (14/9–15/9 and 29/9–30/9) and October (13/10–14/10), and once (24 h) in April (28/4–29/4). The samples were divided with a modified Folsom sample splitter and counted with an inverted plankton microscope. Altogether 148 pooled zooplankton samples consisting of 740 separate samples were taken.

Fish

The fish was collected with a gill net series of 25 m wide and 18 m deep with knot-to-knot mesh sizes of 8, 10, 12, 14, 16, 18, and 20 mm. The sampling station and periods were the same as those described with zooplankton and the catching time was usually 5 hours.

The stomachs were carefully dissected and preserved immediately in 4% formaldehyde in the field and later in the laboratory in 70% alcohol. The amount of food in the stomach was first subjectively estimated by a points method, 0 = empty, 1 = some food in pyloric region, 2 = some food in pyloric and cardic region, 3 = pyloric region full and some in cardic, 4 = full stomach. The contents of the stomach were then sorted into species or higher taxonomic groups and the number of each determined. Food analyses were made separately for the anterior (including the oesophagus) and posterior parts of the stomach. In the anterior part all the specimens were calculated but in the posterior part, when a large number of plankton animals (over about 2 000) had been consumed, the total number was calculated using a modified Stempel-Henzen pipette. Usually four subsamples were taken from each stomach and pooled. The number of animals in these pooled samples was usually 300–600 and in a full stomach about 5 000–8 000 depending on the species consumed and the size of the fish. Large, rare food animals such as *Chaoborus*, chironomid larvae and pupae were counted without subsampling. The number of food animals in the anterior part of the stomach was about 10–20% of the content of the whole stomach. The precision of the dividing and counting of the samples was tested with four replicate counts and five different stomach samples. The coefficient of variation (standard deviation/mean) for the most abundant food species *(Daphnia, Bosmina, Heterocope, Cyclops)* was usually in the range 15–40% (mean 32.8%) but with low numbers of individuals the range was 35–80%.

The percent composition of each species or higher taxonomic group was calculated for both zooplankton and stomach content samples of each depth and times of the day. Rotifers and Copepod nauplius are not included in the figures as they are very common in zooplankton but only occasionally consumed by cisco. In order to test for the selection of various zooplankton by cisco, the percent composition data was used to calculate Ivlev's (1961) electivity index (E) defined as $E = (r_i - p_i) \cdot (r_i + p_t)^{-1}$, where E is the electivity index, r_i is the proportion of the i_{th} food item in the stomach contents, and p_i is the proportion of the same food item in the lake. The values of E range from −1 to +1. An E value of 0 indicates the absence of selective feeding. As E approaches +1, positive selection of a particular food source from the habitat is indicated. As E approaches −1, negative selection of food item is indicated.

A total of 968 fish were collected ranging from 8.2 to 23.4 cm in total length. For food analyses the fish were divided into two size classes: over (two years and older) and under (one summer and one-year old) 16 cm. The number of stomachs analyzed in the former group was 240 and in the latter 88. The percent of empty stomachs was 8.4.

Results

Vertical distribution of cisco

The vertical distribution of cisco in different months and diel periods (Figs. 3–4) based on gill net catches of 866 cisco.

Under the ice (February, March, April) cisco were widely dispersed from surface to bottom. After ice had melted in May, cisco became progressively concentrated near the surface. In June almost all (96%) the fish were in the upper epilimnion. In July the fish were still congregated near the surface although they were more widely dispersed. From August to October cisco again was widely distributed from surface to bottom.

The distribution of fish changed also in diel periods. In June there were no fish in the morning catch. In July, August, September and October the number of fish during the daytime (06.00–18.00) was much less than in other diel periods. Cisco probably migrated away and/or were more inactive during the daytime. A proof of the latter is the fairly high number of empty or almost empty stomachs in that period.

The migration of cisco appears to be dependent on the age and size of the fish. Young cisco (<16 cm) were found to inhabit more often the uppermost layers of water, while older cisco preferred deeper and colder water.

Food composition in different seasons

Figure 2 illustrates the zooplankton composition (column 1) and the type of food eaten each month by the sampled population of cisco under (column 2) and over (column 3) 16 cm.

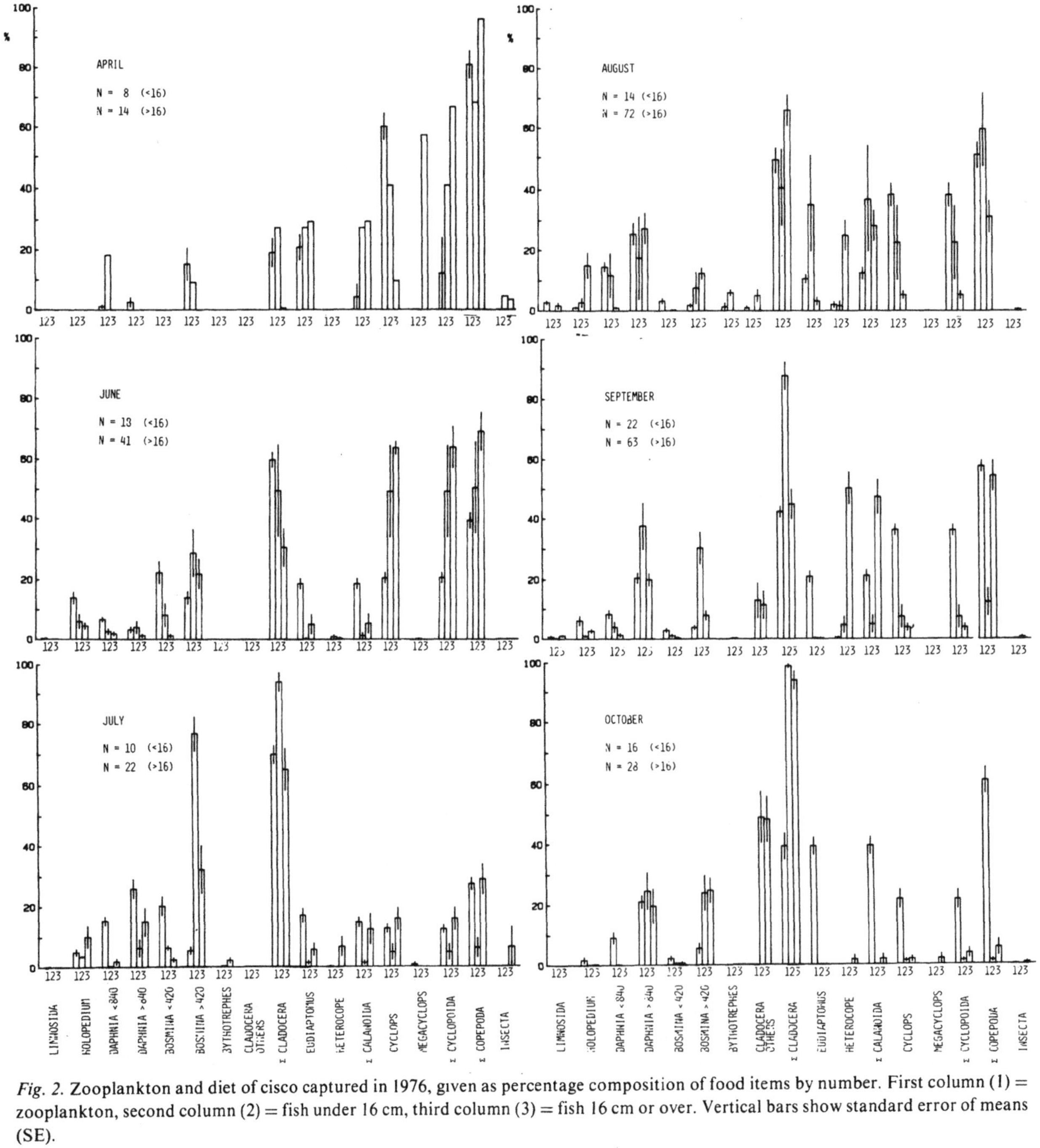

Fig. 2. Zooplankton and diet of cisco captured in 1976, given as percentage composition of food items by number. First column (1) = zooplankton, second column (2) = fish under 16 cm, third column (3) = fish 16 cm or over. Vertical bars show standard error of means (SE).

Some of the names (Figs. 2–5) mean a group of species and the name is given according to the most abundant species. So e.g. *Limnosida* includes *Limnosida frontosa, Diaphonosoma brachyrum* and *Sida crystallina, Daphnia* includes *D. longiremis* and *D. cristata, Bosmina* includes *B. coregoni* and *B. longirostris,* Cladocera others includes those species which belong only occasionally to a diet e.g. *Alona affinis, Alonella nana, Acoperus, Chydorus, Ceriodaphnia, Leptodora kindtii* and Clodocera ephippium, *Cyclops* includes *C. scutifer, Mesocyclops leuckarti* and *Thermocyclops oithonoides, Megacyclops* includes *M. viridis* and *M. gigas,* Insecta includes *Chaoborus flavicans,* Chironomidae larvae and pupae, Hydracarina and terrestrial insects.

Seasonal changes in the diet are the result of the availability of the prey as well as the consequence of activity and behaviour of the prey and of the predator. With minor exceptions, cisco in Lake Suomunjärvi fed on cladocerans and copepods. Of the total number of food items eaten under 16 cm cisco, copepods made up 68% in April; cladocerans 50–59% in June and August; cladocerans 87–96% in July, September and October. *Eudiaptomus gracilis* and *Cyclops scutifer* were the most abundant prey consumed in April; *Bosmina, Cyclops scutifer* and *Mesocyclops leuckarti* in June; *Bosmina* in July; *Daphnia* and *Eudiaptomus gracilis* in August; *Daphnia, Bosmina* and Cladocera ephippium in September and October.

In the size class of 16 cm or over, copepods made up 69–95% in April and June; cladocerans 64–66% in July and August; copepods 55% in September; and cladocerans 94% in October. The dominant species in April were *Megacyclops viridis, Mega-*

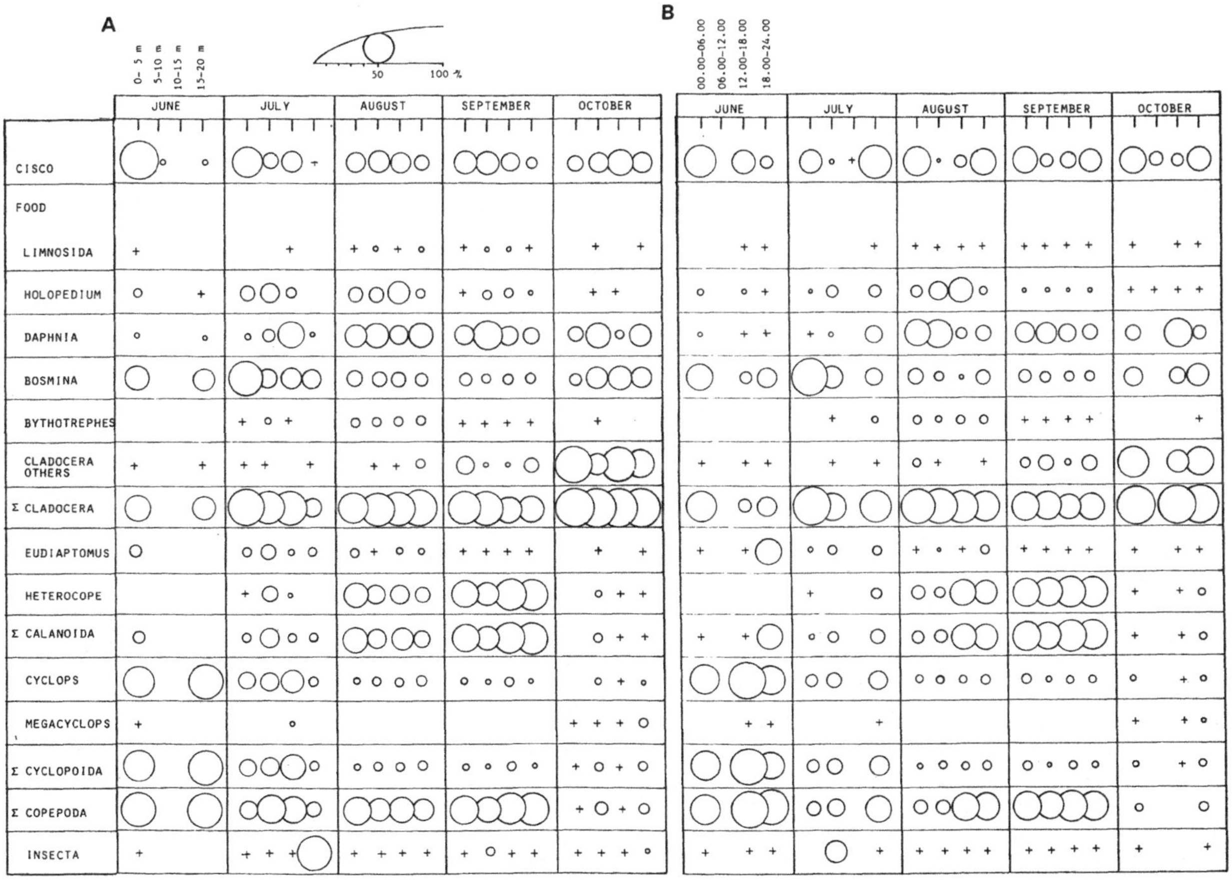

Fig. 3. The diet of cisco 16 cm or over, given as percentage composition of food items by number and distribution (%) of cisco plotted for each depth (A) and diel period (B) for each month.

cyclops gigas and *Eudiaptomus gracilis; Cyclops scutifer* and *Mesocyclops leuckarti* in June; *Daphnia* and *Bosmina* in July; *Daphnia* and *Heterocope appendiculata* in August and September; and *Daphnia, Bosmina* and Cladocera ephippium in October. Large cladocerans like *Bythotrephes longimanus* and *Leptodora kindtii* were strongly selected, but had quantitative importance only in July and August. *Chaoborus flavicans* was most frequent as food in July and Chironomidae in September. The difference in food selection between small and large *Daphnia* and *Bosmina* was very clear. From June to October the share of small *Daphnia* (<840 μm) in cisco diet ranged between 1–12% and large *Daphnia* (≥840 μm) between 3–38%, and Bosmina 1–8% (<420 μm) and 7–78% (≥420 μm), respectively.

Food composition at different depths and diel periods

The composition of zooplankton in a diet of cisco in the lake was different (Figs. 3–4). The distribution of zooplankton was very similar at different depths and diel periods. Diel vertical movements were only found exceptionally.

The composition of cisco diet changed at different depths and times of the day. In June *Eudiaptomus gracilis* were consumed only at a depth of 0–5 m (10%) and in the evening (36%), although it was found in zooplankton from surface to bottom. In July *Daphnia* was most common at depths of 10–15 m (32%) and in the evening (22%). *Bosmina* was most common at the depth of 0–5 m (60%) and during the night (84%). *Chaoborus flavicans* was consumed only near the bottom (63%) and in the

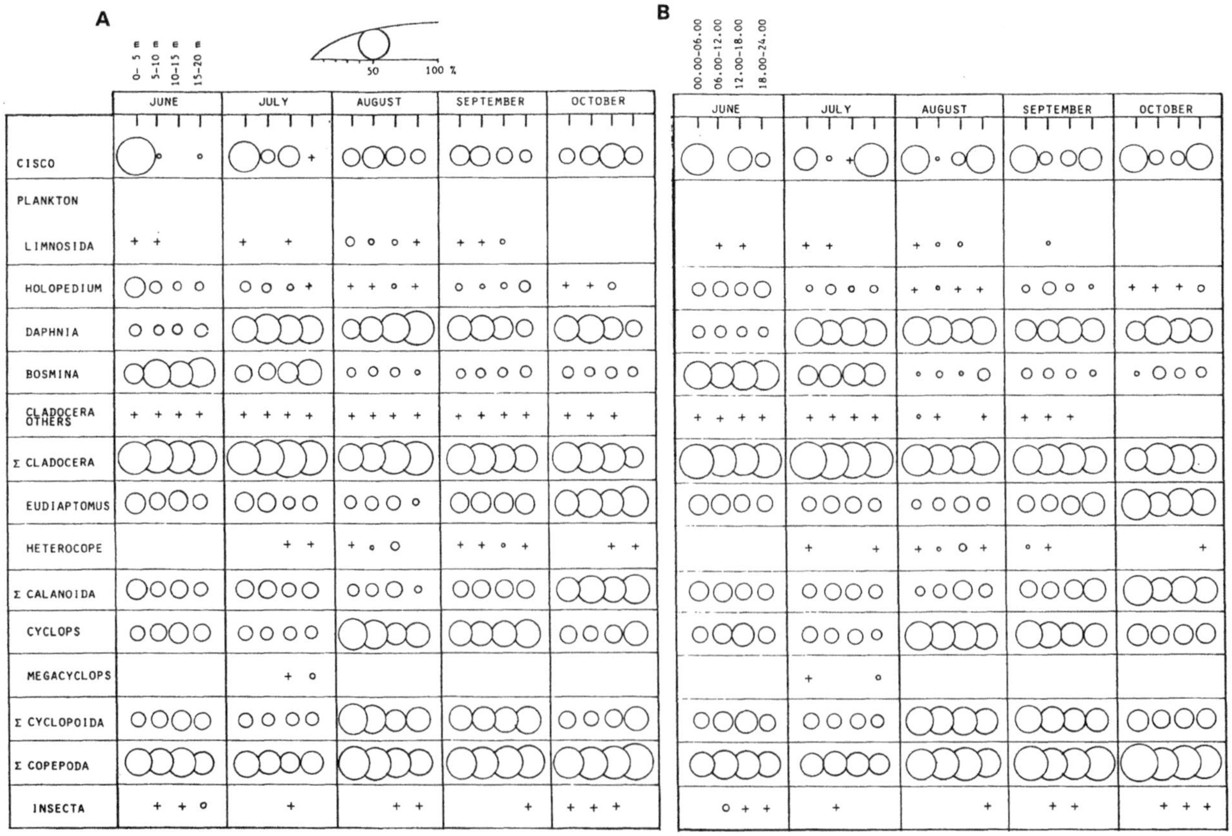

Fig. 4. Percentage composition of zooplankton items by number and distribution (%) of cisco plotted for each depth (A) and diel period (B) for each month.

morning (32%). In August *Holopedium gibberum* and *Heterocope appendiculata* were strongly selected. They were most abundant in the afternoon (35 and 40%), *Holopedium* at the depth of 10–15 m (28%) and *Heterocope* at the depth of 0–5 m (33%). *Daphnia* were consumed in August during the night and early in the morning (45%). In September diurnal differences in a diet were fairly small. *Daphnia* was consumed more at the depth of 5–10 m (44%) and *Heterocope appendiculata* at the depth of 10–20 m (56%). In October many of the stomachs were almost empty and therefore the data gives only an indication of the diet. Cladocera ephippium was very strongly selected in every depth and time of the day in October.

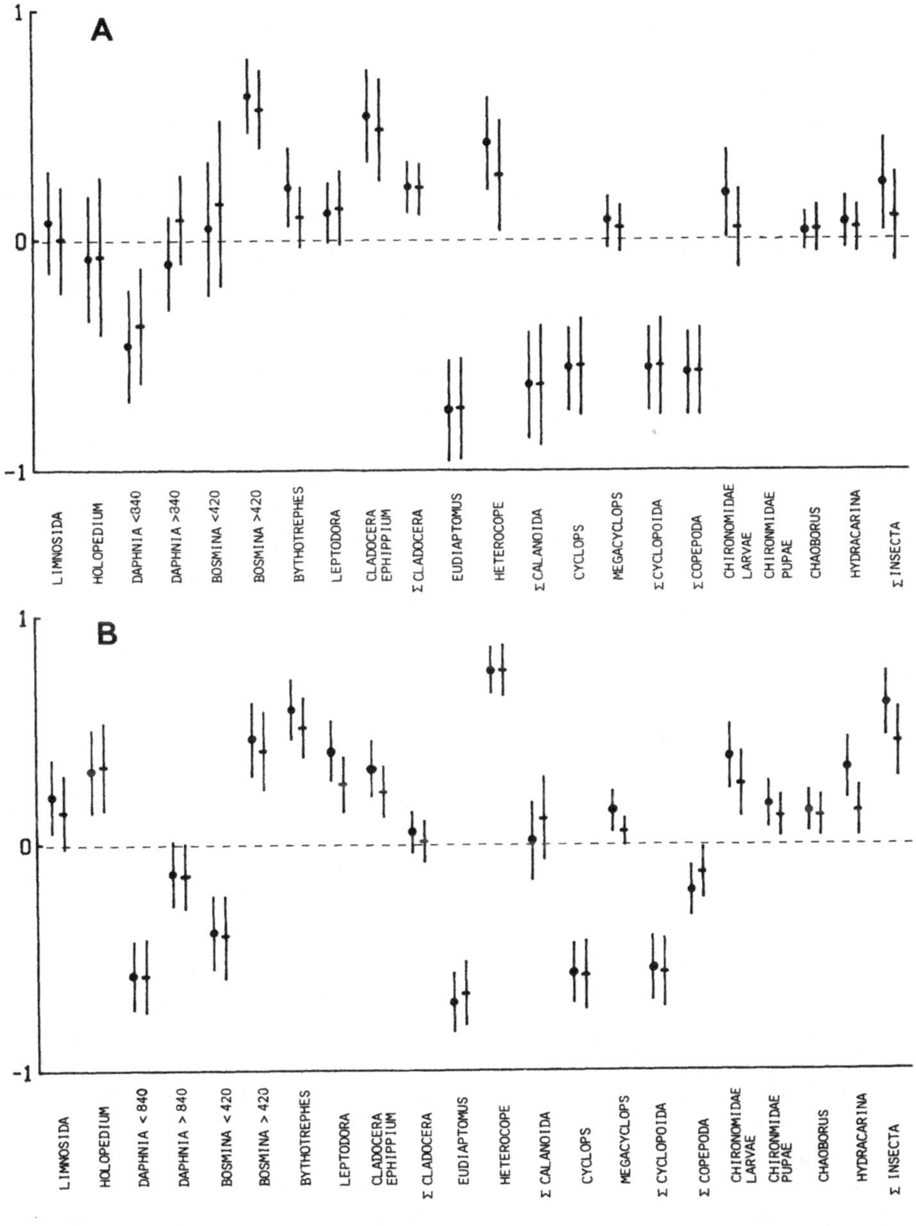

Fig. 5. Electivity indices for zooplankton eaten by 88 cisco under 16 cm (A) and 240 cisco 16 cm or over (B) in 1976. ● = whole stomach, – = anterior part of the stomach. Vertical bars show 95% confidence limits.

Selection of zooplankton

In the size class under 16 cm Ivlev's index (E) shows a preference for large *Bosmina, Bythotrephes longimanus, Leptodora kindtii,* Cladocera ephippium, *Heterocope appendiculata* and chironomid larvae. No preference was noted for *Limnosida frontosa, Holopedium gibberum,* large *Daphnia, Megacyclops, Chaoborus flavicans* and Hydracarina. The species avoided were *Diaphanosoma brachyrum,* small *Daphnia, Eudiaptomus gracilis, Cyclops scutifer* and *Thermocyclops oithonoides.*

In the size class 16 cm or over cisco preferred *Limnosida frontosa, Holopedium gibberum,* large *Bosmina, Bythotrephes longimanus, Leptodora kindtii,* Cladocera ephippium, *Heterocope appendiculata, Megacyclops,* all the insects, and *Daphnia* small *Bosmina, Eudiaptomus gracilis, Cyclops scutifer, Thermocyclops oithonoides* were avoided.

The preference or avoidance of zooplankters changed in different months (Fig. 2).

Comparison of the indices between the anterior part and the whole stomach showed very small differences and were not statistically significant ($p > 0.05$) although the values of selection were usually smaller in the anterior part of the stomach. Comparison between the anterior and posterior part of stomach gave a clearer difference although the difference was significant ($p < 0.05$) only in some cases (small *Bosmina, Bythotrephes longimanus, Leptodora kindtii, Hydracarina, Chaoborus flavicans*). Probably these small differences follow from the fact that the distribution of zooplankton was very similar in each period irrespective of the depth or diel period.

Discussion

It is a well-known fact that cisco display some sort of seasonal and diel vertical and horizontal migrations. However, clear diel vertical migrations were not found in Lake Suomunjärvi in 1976. Reasons for that are probably the poor stratification of the lake, a very small volume of hypolimnion, especially in July and August (0.8%), and that the distribution of zooplankton was very similar at different depths and diel periods and diel vertical movements of zooplankton were only found rarely *(Eudiaptomus gracilis, Cyclops scutifer).*

Usually diel vertical migrations of fish are connected with the formation of metalimnion. The fish try to optimize feeding and avoid unfavourable conditions. According to Brett (1967, 1971) and Fry (1971) the swimming activity of fish is highly temperature-dependent. Diel migrations are an adaptation for saving energy by feeding in the upper parts of the waters at dusk and then migrating down to colder water to digest the food in a temperature zone which is optimal for the species in question. However, cisco has been found to make vertical migrations also under homothermal conditions and changes in illumination and diel vertical migrations of zooplankton has been suggested as explanation for this (Järvi 1919; Airaksinen 1967; Northcote & Rundberg 1970; Piipponen 1977; Hakkari 1978; Hamrin 1979).

For cold stenothermal species like cisco with a high basal metabolism at relatively low temperatures, the oxygen concentration of the water can become a limiting factor at higher temperatures. According to Walczak (1953) and Hamrin (1979) temperatures around 16 °C seem to be an upper limit for the occurrence of cisco in the epilimnion.

Migration to a rich feeding habitat increases growth rate, survival and fecundity, and reduces competition for food.

The food of cisco in Lake Suomunjärvi is dominated by cladocerans and copepods. Cisco were usually highly selective of larger species of cladocerans and copepods and a small cladoceran, *Bosmina coregoni,* as found also by Marciak (1962), Vallin (1969), Hakkari (1972, 1978) and Hamrin (1979). The strong reproductive capability of *B. coregoni* allows it to withstand the predation (Hakkari 1978). The ability of cisco to use *B. coregoni* and to change the zooplankton community from large to small forms by predation could be one important factor making cisco superior to other pelagic fish.

Fish feeding mechanisms determine the relative roles of planktivore size selectivity and zooplankton escape ability in influencing fish feeding selectivity (Drenner & McComas 1980). However fish size selectivity and zooplankton escape ability are not the only factors governing selective feeding. Calculations of fish electivity indices have often shown that fish can select small cladoceran, e.g. *Bosmina* over the other cladocerans, even though *Bosmina* has a relatively small body size and small eye-pigmentation diameter (Zaret & Kerfoot 1975;

Zaret 1980). Therefore prey body-size and prey visibility are not sufficient to explain planktivore selection of prey items. Zaret (1980) has proposed that prey species in motion are most vulnerable to fish predation and he suggests that planktivores can facilitate prey-capture success by learning, to recognize prey items according to their motion. On the other hand Zaret (1972) has shown that the size of pigmented area is important for the selection of zooplankton with transparent bodies. According to Janssen (1980) cisco have both selective and nonselective (in the dark) methods for feeding on zooplankton.

An investigation based on comparing the stomach contents of the fish with the composition of the zooplankton is subject to criticism, because fish and zooplankton form patches both vertically and horizontally, and since fish concentrations and zooplankton concentrations need not overlap, what a plankton net collects may not be what is available to the fish. A proof of this in Lake Suomunjärvi is the abundant presentation of e.g. *Leptodora kindtii, Bythotrephes longimanus, Chaoborus flavicans* in the catches of tow nets (modified Gulf V sampler) although there were very few or none in the usual zooplankton samples. The method is also subject to the criticism that different zooplankters digest at different rates thus biasing counts of prey in stomachs. *Daphnia* has been found to be digested most rapidly (Gannon 1976) and the apparent negative selectivity may be erroneous due to rapid rate of digestion.

Jacobs (1974) has criticized Ivlev's electivity index and showed that the index is not an appropriate measure of prey selection since it depends on the abundance of prey and not just on the predators preference. However, Paloheimo (1979) has shown that the modification proposed by Jacobs (1974) is similarly affected by the abundance. Paloheimo (1979) has suggested that an index of preference that is indeed independent of the prey abundance can be based on standardized forage ratios, standardized so that the forage ratios for the different prey species sum to one.

Acknowledgements

My thanks to the staff of the Karelian Institute, Section of Ecology, for valuable assistance, to R. Latja, who analyzed zooplankton samples, and to Mr Kenneth Meaney, M.A. for revision of the English text. This work was financially supported by the Academy of Finland.

References

Airaksinen, K., 1967. Varmavirran muikku (The vendace (Coregonus albula L.) of Varmavirta in the northern part of Lake Saimaa (SE-Finland)). Suom. Kalatal. 30: 1-32 (in Finnish, with English summary).

Brett, J. R., 1967. Swimming performance of sockeye salmon (Oncorhynchus nerka) in relation to fatigue time and temperature. J. Fish. Res. Bd Can. 24: 1731-1741.

Brett, J. R., 1971. Energetic response of salmon to temperature. A study of some thermal relations on the physiology and freshwater ecology of Sockeye salmon (Oncorhynchus nerka). Am. Zool. 11: 99-113.

Brooks, J. L. & Dodson, S. I., 1965. Predation, body size, and composition of plankton. Science 150: 28-35.

Dembinski, W., 1971. Vertical distribution of vendace (Coregonus albula L.) and other pelagic fish species in some Polish lakes. J. Fish. Biol. 3: 341-357.

Drenner, R. W. & McComas, S. R., 1980. The roles of zooplankter escape ability and size selectivity in the selective feeding and impact of planktivorous fish. In: Kerfoot, W. C. (ed.). Evolution and Ecology of Zooplankton Communities, pp. 587-593. University Press of New England, Lond.

Engel, S., 1976. Food habits and prey selection of coho salmon (Oncorhynchus kisutch) and cisco (Coregonus artedii) in relation to zooplankton dynamics in Pallette Lake, Wisconsin. Trans. Am. Fish. Soc. 105: 607-614.

Engel, S. & Magnuson, J. J., 1976. Vertical and horizontal distributions of coho salmon (Oncorhynchus kisutch), yellow perch (Perca flavescens), and cisco (Coregonus artedii) in Pallette Lake, Wisconsin. J. Fish. Res. Bd Can. 33: 2710-2715.

Fry, F. E. J., 1971. The effect of environmental factors on the physiology of fish. In: Hoar, W. S. & Randall, D. J. (eds.). Fish Physiology 6, pp. 1-98. Academic Press, New York.

Gannon, J. E., 1976. The effects of differential digestion rates of zooplankton by alewife, Alosa pseudoharengus, on determinations of selective feeding. Trans. Am. Fish. Soc. 105: 89-95.

Gliwicz, S. M. & Prejs, A., 1977. Can planktivorous fish keep in check planktonic crustacean populations? A test of size-efficiency hypothesis in typical Polish lakes. Ekol. pol. 25: 567-591.

Hakkari, L., 1972. Konneveden eläinplanktonin merkityksestä muikun ravintona (The role of zooplankton as food for vendace (Coregonus albula) in Lake Konnevesi, Central Finland). Suom. Kalatal. 46: 21-28 (in Finnish, with English summary).

Hakkari, L., 1978. On the productivity and ecology of zooplankton and its role as food for fish in some lakes in central Finland. Biol. Res. Rep. Univ. Jyväskylä 4: 3-87.

Hammar, S., 1968. Näringsekologi hos fisk i Lilla Ullevifjärden. Inform. Inst. Freshwat. Res. Drottningholm 10, 16 pp. (in Swedish).

Hamrin, S. F., 1979. Populationsdynamik, vertikalfördelning och fodoval hos siklöja, Coregonus albula L. i sydsvenska sjöar (Population dynamics, vertical distribution and food selection of ciscoe (Coregonus albula L.) in south Swedish lakes). Inst. Limnol., Univ. Lund, Lund, 195 pp. (in Swedish, with English summary).

Hrbáček, J., 1962. Species composition and the amount of zooplankton in relation to the fish stock. Rozpr. čsl. Akad. Věd Řada MPV 72: 1-116.

Ivlev, V. S., 1961. Experimental ecology of the feeding of fishes. Yale University Press, New Haven, 302 pp.

Jacobs, J., 1974. Quantitative measurement of food selection: a modification of the forage ratio and Ivlev's electivity index. Oecologia 14: 413-417.

Janssen, J., 1980. Alewives (Alosa pseudoharengus) and ciscoes (Coregonus artedii) as selective and non-selective planktivores. In: Kerfoot, W. C. (ed.). Evolution and Ecology of Zooplankton Communities, pp. 580-586. Univ. Press of New England, Lond.

Järvi, T. H., 1919. Muikku ja muikkukannat 1. Keitele. Suom. Kalatal. 5: 1-284 (in Finnish).

Latja, R., 1978. Suomunjärven muikun ravintolajien biologiasta vuonna 1976. In: Meriläinen, J. (ed.). Kalantuotantoon vaikuttavat tekijät luonnonvaraisessa dys-oligotrofisessa järvessä (Suomunjärvi, Lieksa). Loppuraportti 31.12.1978. pp. 129-225. Joensuun korkeakoulu, Karjalan tutkimuslaitos, Joensuu 1978 (manuscript, in Finnish).

Maitland, P. S., 1968. Echo sounding observations on the Lochmaben vendace, Coregonus vandesius Richardson. Trans. J. Proc. Dumfries. Galloway nat. Hist. antiq. Soc. 44: 29-46.

Marciak, Z., 1962. Sezonowe zmiany w odzywianin sie i wzroscie sielawy (Coregonus albula L.) z jeziora Pluszne (Seasonal variations in feeding habits and growth of Coregonus albula L. in Pluszne Lake). Roczn. Nauk. roln. B 81: 335-357 (in Polish, with English summary).

Nilsson, N.-A., 1979. Food and habitat of the fish community of the offshore region of Lake Vänern, Sweden. Inst. Freshwat. Res. Drottningholm, Rep. 58: 126-139.

Northcote, T. G., 1978. Migratory strategies and production in freshwater fishes. In: Gerking, S. D. (ed.). Ecology of freshwater fish production, pp. 326-359. Blackwell Scientific Publications, Oxford.

Northcote, T. G. & Rundberg, H., 1971. Spatial distribution of pelagic fish in Lambarfjärden (Mälaren, Sweden) with particular reference to interaction between Coregonus albula and Osmerus eperlanus. Inst. Freshwat. Res. Drottningholm, Rep. 50: 133-167.

Paloheimo, J. E., 1979. Indices of food preference by a predator. J. Fish. Res. Bd Can. 36: 470-473.

Patalas, K., 1954. Skorupiaki planktonowe jako baza pokarmowa w gospodarce sielawowej na jeziorze Charzykowo (Plankton Crustacea as a food base for the small whitefish (Coregonus albula) population in the Charzykowo lake). Pol. Arch. Hydrobiol. 2: 259-276. (in Polish, with English summary).

Piipponen, M.-L., 1977. Planktonäyriäisten lajisto, biomassa, tuotanto sekä vuorokaudenaikaiset vertikaaliliikkeet suhteessa muikun liikkeisiin Lieksan Suomunjärvessä (Seasonal and diurnal fluctuations of planktonic crustaceans, and vertical migrations of vendace (Coregonus albula L.) - Suomunjärvi, eastern Finland). Joensuun korkeakoulu, Karjalan tutkimuslaitos, julkaisuja 21: 1-25 (in Finnish, with English summary).

Smith, S. H., 1956. Life history of lake herring of Green Bay, Lake Michigan. Fish. Bull. Fish Wildl. Serv. U.S. 109: 1-138.

Stenson, J. A. E., 1972. Fish predation effects on the species composition of the zooplankton community in eight small forest lakes. Inst. Freshwat. Res. Drottningholm, Rep. 52: 132-148.

Szypula, J., 1970. Odzywianie sie nayblen sielawy w jezioreze Legińskim i Wydryńskim (Feeding habits of Coregonus albula L. in Leginskie and Wydrynskie lake). Roczn. Nauk. roln. H 92: 61-84 (in Polish, with English summary).

Vallin, S., 1969. Siklöjans näringsbiologi i Lambarfjärden i Mälaren. Inf. Sötvattenslaboratoriet Drottningholm 7, 57 pp. (in Swedish).

Viljanen, M., 1972. Kalojen ravinnosta ja kasvusta Pääjärvessä (EH). Helsingin yliopisto, eläintieteen laitos, 98 pp. (manuscript, in Finnish).

Viljanen, M., 1974. Pääjärven kalasto. (Fishes of Pääjärvi). Luonnon Tutk. 78: 174-180 (in Finnish, with English summary).

Viljanen, M. & Holopainen, I. J., 1982. Population density of perch (Perca fluviatilis L.) at egg, larval and adult stages in the dys-oligotrophic Lake Suomunjärvi, Finland. Ann. zool. fenn. 19: 39-46.

Walczak, J., 1953. Sielawa (Coregonus albula L.) kilku jezior Pomorza Zachodniego (Movements and distribution of the small whitefish (Coregonus albula L.) shoals in Lake Zachodniego). Roczn. Nauk. roln. B 67: 40-51 (in Polish, with English summary).

Zaret, T. M., 1972. Predators, invisible prey, and the nature of polymorphism in the cladocera (Class Crustacea). Limnol. Oceanogr. 17: 171-184.

Zaret, T. M., 1980. The effect of prey motion on planktivore choice. In: Kerfoot, W. C. (ed.). Evolution and Ecology of Zooplankton Communities, pp. 594-603. Univ. Press of New England, Lond.

Zaret, T. M. & Kerfoot, W. C., 1975. Fish predation on Bosmina longirostris: visibility selection versus body-size selection. Ecology 56: 232-237.

Differences in growth of perch (*Perca fluviatilis* L.) in two small forest lakes

Martti Rask
Lammi Biological Station, University of Helsinki, SF-16900 Lammi, Finland

Keywords: forest water ecosystems, perch, growth, forest lake, population density, bottom fauna, zooplankton

Abstract

Growth patterns and food composition of perch, *Perca fluviatilis* L., was studied in two small forest lake populations in southern Finland. Size and morphometry of the lakes and physical and chemical properties of water are similar. There is a clear difference in the growth rates of perch between the two lakes. The difference in growth is highly significant in all age groups. In the first lake there is a perch population of 2 000 (1 750 ind · ha^{-1}) adult fishes. In the second lake there is a small population of pike, that keeps the perch population down: 200 adult perch (530 ind · ha^{-1}). The main food items of perch are crustacean zooplankton, *Asellus aquaticus* L. and Trichoptera larvae in the first lake and zooplankton, Odonata larvae, Ephemeroptera larvae and Heteroptera in the second.

It is concluded that the main reason for the growth difference of studied perch populations is the different population density. There are also differences in species composition of bottom fauna of the lakes, maybe owing to the floating Sphagnum peat moss vegetation in the second lake. This can also affect the growth difference between the two populations of perch.

Introduction

The perch, *Perca fluviatilis* L., is widely distributed from sea level to over 1 000 m altitude in Eurasia. It has been introduced into Australia and New Zealand (Jellyman 1980) and South Africa. According to Thorpe (1977) it seems prudent to regard Eurasian perch and American yellow perch, *Perca flavescens* Mitchell, as functionally the same.

Perch is adapted to different inland watercourses (Hartmann 1975) as well as Baltic archipelagoes (Neuman 1976). In many small forest lakes perch is the only fish species (Nyberg 1976). In several small forest lakes the fish fauna is represented by only two sizes of perch: small ones feeding on zooplankton and benthos near the shore and larger ones cannibalizing the small ones. Perch is often a source of food for larger predators (Popova & Sytina 1977). In Finnish Baltic archipelago perch can be predator during its first summer, 5 cm long, feeding on gobies (Koli *et al.* 1978). Food items vary among populations and are probably largely dependent on availability of the organisms (Elrod *et al.* 1981).

The growth of perch varies widely between the watercourses. Tesch (1955) grouped the growth rate of perch into 5 categories from very poor, all individuals smaller than 16 cm, to very good, age group 2 over 20 cm.

Small polyhumic forest lakes have some typical features that affect the ecology of perch: small area, dark water, scarcity of nutrients which affect the productivity of the lake and poor oxygen conditions in the hypolimnion. There are few appropriate food species and the perch is often the only species of fish. Since the ecosystem of these lakes is often relatively simple, they present good opportunities to study the interactions between a species and its environment, such as, in this work, the factors affecting the growth of perch.

Hydrobiologia 101, 139–144 (1983).

The lakes

The study lakes are small, polyhumic forest lakes in southern Finland (cf. Arvola, in press) with areas 1.1 and 0.4 hectares and the oxic epilimnion is on average two meters thick.

In the first lake, Lake Horkkajärvi, there is a small population (200 transplanted in 1978) of whitefish, *Coregonus peled* Gmelin, the mean length of which was about 30 cm after the growing season of 1980. No reproduction of whitefish occurs in this lake.

In the second lake, Lake Nimetön, there is a small population of pike, *Esox lucius* L., besides the perch. In the spring of 1979 and 1980, 5–10 individuals ofer 20 cm total length (max. 42.6 cm) were captured and mostly released. Smaller pike (10–20 cm) occurred in both summers, indicating reproductive success in both years.

Material and methods

Fish samples were taken in the ice-free period of 1979 and 1980 at intervals of two weeks from both lakes. Traps were used with a mesh size of 1 cm. Such a trap takes perch of more than 8.5 cm total length. Because of the small size of the lakes, no special fishing places were used. In Lake Horkkajärvi fish were captured along the whole shoreline. In Lake Nimetön there are only two possible places for traps, the rest of the shore is steep with no oxygen at the bottom.

From every catch samples of 10–20 perch (300 in total) from Lake Horkkajärvi and 5–10 (95 in total) from Lake Nimetön were taken for determination of age, growth and food. The length distributions of the samples corresponded to those of catches. Total length (mm) and weight (g, wet weight) were measured, opercular bone for age and growth and stomach for food analyses were collected.

Plots of fish length against opercular bone length were drawn separately for both populations and the procedure given by Le Cren (1947) was followed in determining the growth. In food analyses a volumetric points method was used with a scale of 0–24 points (0 = empty, 12 = half full, 24 = full). The size of populations was estimated in spring 1979 by mark and recapture. The left pectoral fin was cut and length of the marked individuals was measured for length distribution.

Results

1 480 perch were captured and marked from Lake Horkkajärvi in May 1979. Recapture after 3 days gave an estimation of population number 1943 ± 45 (95% conf. limits). The recapture was repeated three days later and the corresponding numbers were 1913 ± 54 (95% conf. limits). The density of perch was thus 1 750 individuals per hectare. No mortality due to marking was observed, the relation of marked fishes to unmarked remained the same during the whole growing season. Later in the summer of 1979 the young of the year class 1978 (1+) grew to catchable size and became the largest length group in the population.

From Lake Nimetön 103 perch were marked, and recaptured in relation 1:1 to unmarked. The estimate of population number 212 ± 53 (95% conf. limits) is the mean of four separate recaptures. The population density was 530 individuals per hectare. Sumari (1971) has given a density value of 700 individuals per hectare as the mean of 22 ponds in Finland. In Lake Suomunjärvi, eastern Finland, the density of perch population is 248 individuals per hectare (Viljanen 1978). There is a clear difference between the size- and age distributions of the studied populations (Fig. 1).

The growth of perch is faster in Lake Nimetön

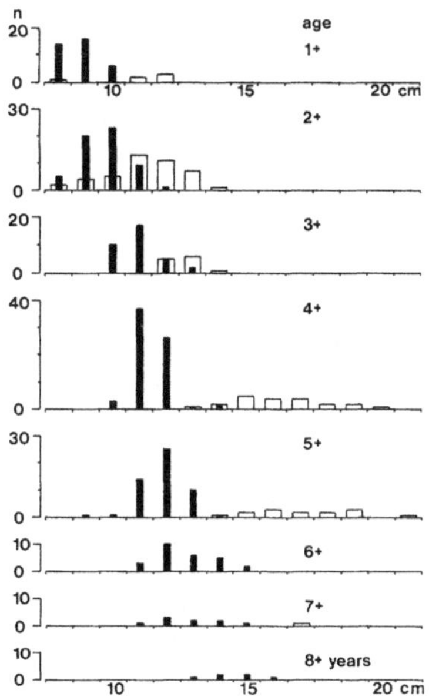

Fig. 1. The age and length distributions of perch. Dark columns Lake Horkkajärvi; white columns Lake Nimetön.

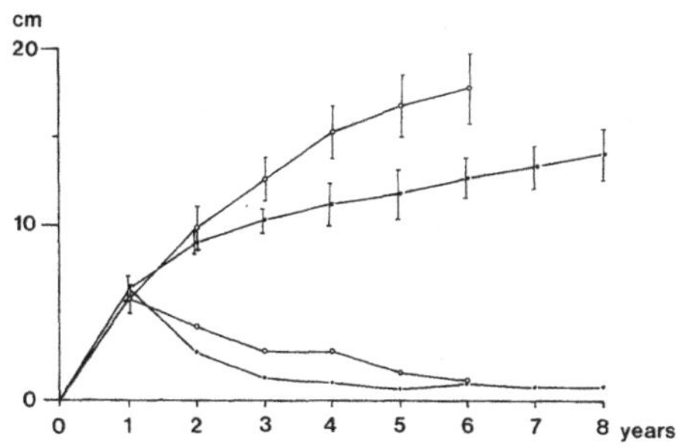

Fig. 2. The growth of perch in Lake Horkkajärvi (dots) and in Lake Nimetön (open circles) and annual length increments.

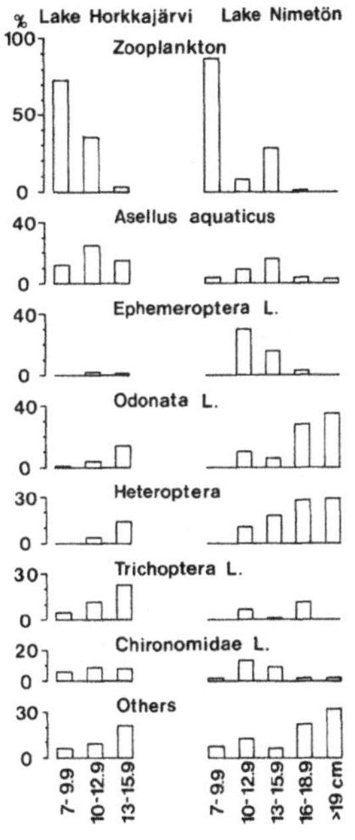

Fig. 3. The food of perch of different length in studied lakes. The values are percent of the total amount of fullness points per length group.

than in Lake Horkkajärvi except in the first year of life (Fig. 2). The mean weight of 3-year old individuals is 11 g in Lake Horkkajärvi and 21 g in Lake Nimetön, and of 6-year old 22 and 62 g respectively. The difference in growth was highly significant in all age groups (Student's *t* test, $p < 0.001$). The growth rate is low in both lakes in comparison to several European lakes (Deelder 1951). In growth category (Tesch 1955) the growth of perch is very poor in Lake Horkkajärvi (all fishes smaller than 16 cm) and poor in Lake Nimetön (the mean length of age group 3 less than 16 cm). The interpretation of fish age from the opercular bone of perch from Lake Horkkajärvi was difficult because of many false annual marks which seem to be related to the slow growth rate. This problem did not exist in Lake Nimetön. In both lakes males grow faster during the first two years. After that the growth rate of females is higher. Alm (1959) and Nyberg (1976) have observed the same feature of perch growth. Males attain sexual maturity at the age of two years having a length then of 8–10 cm. Females mature on average a year later, 10–11 cm long in Lake Horkkajärvi and 12–13 cm in Lake Nimetön.

Production and biomass of perch was estimated from population size and growth rate. In 1979 in Lake Horkkajärvi they were 7.9 and 25.0 kg (0.7 and 2.3 g · m^{-2}) and in Lake Nimetön 2.6 and 6.6 kg (0.6 and 1.7 g · m^{-2}, ww.). Corresponding P/B relations were 0.3 and 0.4. In these values the effect of age groups 0+ and 1+ is excluded. The values agree with those reported from Swedish forest lakes (Nyberg 1976). The production and biomass of perch in forest lakes vary from year to year owing to different occurrence of the two youngest age groups, which have relatively better growth and production than the older ones. In Lake Horkkajärvi for example in 1979 the production of the age group 1+ was about the same as the total production of 2+ and older.

The mortality of 2+ and older perch in Lake Horkkajärvi was 0.3. Mortality in Lake Vitalampa, Sweden, was 0.3 in age group 2+ and 0.26 in older age groups (Nyberg 1976). Craig *et al.* (1979) have given estimates of the proportion of the adult perch population dying per annum varying between 0.34–0.72 during the period 1949–1975 in Windermere.

The main food items of perch are crustacean zooplankton, *Asellus aquaticus* L. and Trichoptera larvae in Lake Horkkajärvi and zooplankton, Odonata larvae, Ephemeroptera larvae and Heteroptera (mostly Corixidae) in Lake Nimetön. Chironomid larvae have equal significance in both lakes (Fig. 3).

The most important groups in benthos are *Pisidium* sp. *Asellus aquaticus* and Trichoptera larvae in Lake Horkkajärvi and Ephemeroptera larvae (especially during early summer) and Odonata larvae in Lake Nimetön (Hiisivuori, unpublished data).

Little cannibalism occurs; only two cases per lake were observed. Zooplankton is most important for small individuals up to 10–12 cm. Zooplankton has a maximum importance in the diet of perch in Lake

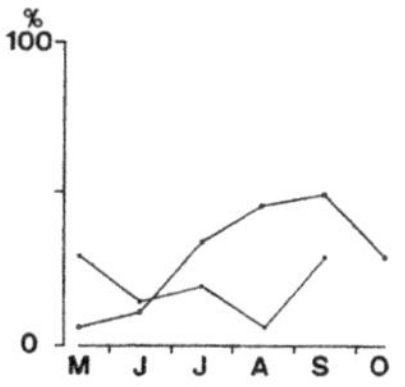

Fig. 4. The share of crustacean zooplankton in the diet of perch in Lake Horkkajärvi (open circles) and Lake Nimetön (dots).

Horkkajärvi in August–September but is less important in the early summer and later autumn. Such a clear variation has not been recorded in Lake Nimetön (Fig. 4). The importance of *Asellus aquaticus* and Ephemeroptera and Chironomid larvae is greatest for fishes 10–16 cm long. The portion of Odonata larvae, Heteroptera and Trichoptera larvae in the diet increases with the size of fish.

In both lakes the food of perch and composition of benthic fauna correspond to each other with a few exceptions. In Lake Horkkajärvi the most abundant species of benthic fauna, *Pisidium*, is not eaten by perch. In Lake Nimetön the importance of Heteroptera as food is greater than one could expect from the composition of the benthic fauna.

Whitefish in Lake Horkkajärvi feed on zooplankton, *Hydracarina* and *Pisidium*. Pike in Lake Nimetön has perch as food.

Discussion

The optimal environmental conditions for perch in lakes are equivalent to those in large, temperate rivers in littoral and sublittoral zone (Kitchell *et al.* 1977). Environmental conditions in small polyhumic forest lakes studied in this work are far from optimal, which can be seen as low growth rates of perch.

The perch population of Lake Horkkajärvi is a stunted one, the population of Lake Nimetön is not. Poor food supply and lack of appropriate prey fish are given as causes of stunted growth by several authors (Alm 1946; Deelder 1951). The latter explanation is not suitable here, because there is no appropriate prey fish population in either of the lakes. The main reason for different growth of perch in these lakes is the different food conditions. This can be caused by different conditions for biological productivity in the lakes, different population densities of perch owing to the existence of pike in Lake Nimetön and different composition and abundance of benthic fauna in the lakes.

The conditions for biological productivity in the lakes are equal (nutrients, oxygen and temperature stratification, etc.). The rate of planktonic primary production is nearly the same (Arvola, in press). So this can not explain the growth difference of perch.

In Lake Horkkajärvi the predation by the perch population on the bottom fauna is stronger than in Lake Nimetön because of its high density (1 750 individuals per hectare). One of the most important food items of 2+ and older perch, *Asellus aquaticus,* has a shift of generation at the beginning of July. When the adult generation of *Asellus* dies, the food supply to perch decreases. The poor availability of Trichoptera larvae at the same time for perch has the same impact. In midsummer only in 1.5 m layer of surface water is there enough oxygen for perch, which makes the littoral zone, the feeding area of perch very narrow. Thus the adult perch must feed on zooplankton more than in the spring or later in the autumn. This is very inefficient because of the feeding habit of perch, 'snipping' one by one, and the high metabolic rate of fish in warm water. Consequently growth is slow or ceases in late summer. Conditions for growth in Lake Horkkajärvi resemble those in Lake Vitalampa in Sweden (Nyberg 1976). The most important food item of whitefish is zooplankton, but the high growth rate of perch of age groups 0+ and 1+ in comparison to those in Lake Nimetön shows that no serious competition of food affecting the growth of perch occurs. Because the optimal growth of perch in these circumstances is unknown, one can not say anything exact about the impact of whitefish on the growth of perch. One can conclude that there is enough food (zooplankton) for the youngest fishes in lakes of that kind, but a severe shortage of food can occur, when the fish, here perch, should change their diet to benthic fauna.

In Lake Nimetön there is no such strong predation on the bottom fauna owing to the smaller density (530 individuals per hectare) of the perch population. Clearly pike predation controls the population density of perch and thus prevents their overgrazing the benthic fauna. The floating Sphagnum peat moss is an inverted bottom which offers a favourable environment for Odonata larvae, Ephemeroptera larvae and Corixidae, which are the most important food items of perch here. Most Ephemeroptera larvae hatch in June–July and are not available for perch in late summer, but according to life cycles of Odonata (Macan 1962) and Corixidae (Pajunen 1977) they are available for perch during the whole growing season. In Lake Nimetön there are enough benthic food organisms throughout the summer. So the portion of zooplankton in the diet of perch is not so significant as in Lake Horkkajärvi. The growth conditions and also the growth, are better during the whole growing season.

References

Alm, G., 1946. Reasons for the occurrence of stunted fish populations with special reference to the perch. Rep. Inst. freshwat. Res. Drottningholm 25, 146 pp.

Alm, G., 1959. Connection between maturity, size and age in fishes. Rep. Inst. freshwat. Res. Drottningholm 40: 5–145.

Arvola, L., in press. Seasonal succession of phytoplankton in two small, polyhumic forest lakes in S. Finland.

Craig, J. F., Kipling, C., Le Cren, E. D. & McCormack, J. C., 1979. Estimates of the numbers, biomass and year-class strengths of perch (Perca fluviatilis L.) in Windermere from 1967 to 1977 and some comparisons with earlier years. J. anim. Ecol. 48: 315–325.

Deelder, C. L., 1951. A contribution to the knowledge of the stunted growth of perch (Perca fluviatilis L.) in Holland. Hydrobiologia 3: 357–378.

Elrod, J. H., Busch, W.-D. N., Griswold, B. L., Schneider, C. P. & Wolfert, D. R., 1981. Food of white perch, rock bass and yellow perch in eastern Lake Ontario. N. Y. Fish Game J. 28: 192–201.

Hartmann, J., 1975. Der Barsch (Perca fluviatilis) im eutrophierten Bodensee. Arch. Hydrobiol. 76: 269–286.

Jellyman, D. J., 1980. Age, growth, and reproduction of perch, Perca fluviatilis L., in Lake Pounui. N. Z. J. mar. freshwat. Res. 14: 391–400.

Kitchell, J. F., Johnson, M. G., Minns, C. G., Loftus, K. H., Greig, L. & Olver, C. H., 1977. Percid habitat: The river analogy. J. Fish. Res. Bd Can. 34: 1936–1940.

Koli, L., Aro, E. & Rask, M., 1978. Tvärminnen ahvenen populaatiotutkimus. Loppuraportti Suomen Akatemian luonnontieteelliselle toimikunnalle. A preliminary research report (in Finnish). Helsinki. 12 pp.

Le Cren, E. D., 1947. The determination of the age and growth of the perch (Perca fluviatilis) from the opercular bone. J. anim. Ecol. 16: 188–204.

Macan, T. T., 1962. Ecology of aquatic insects. Ann. Rev. Ent. 7: 261–288.

Neuman, E., 1976. The growth and year-class strength of perch (Perca fluviatilis L.) in some Baltic archipelagoes, with special reference to temperature. Rep. Inst. freshwat. Res. Drottningholm 55: 51–70.

Nyberg, P., 1976. Production and food consumption of perch in two Swedish forest lakes. Scripta Limnol. Ups. 421, Klotenprojektet Rap. 6: 1–97.

Pajunen, V. I., 1977. Population structure in rock-pool corixids (Hemiptera, Corixidae) during the reproductive season. Ann. zool. fenn. 14: 26–47.

Popova, O. A. & Sytina, L. A., 1977. Food and feeding relations of Eurasian perch (Perca fluviatilis) and pikeperch (Stizostedion lucioperca) in various waters of the USSR. J. Fish. Res. Bd Can. 34: 1559–1570.

Sumari, O., 1971. Structure of the perch populations of some ponds in Finland. Ann. zool. fenn. 8: 406–421.

Tesch, F.-W., 1955. Das Wachstum des Barsches (Perca fluviatilis L.) in verschiedenen Gewässern. Z. Fisch. 4: 321–420.

Thorpe, J. E., 1977. Morphology, physiology, behaviour, and ecology of Perca fluviatilis L. and P. flavescens Mitchell. J. Fish. Res. Bd Can. 34: 1504–1514.

Viljanen, M., 1978. Population studies of vendace (Coregonus albula L.) and perch (Perca fluviatilis L.) in a mesohumic oligotrophic lake. Verh. int. Ver. Limnol. 20: 2103–2110.

Ecological effects of lime treatment of acidified lakes and rivers in Sweden

Fritz Eriksson[1], Einar Hörnström[2], Per Mossberg[1] & Per Nyberg[3]
[1] *Institute of Limnology, Box 557, S-751 22 Uppsala, Sweden*
[2] *National Environment Protection Board, Fack, S-171 20 Solna, Sweden*
[3] *Institute of Freshwater Research, S-170 11 Drottningholm, Sweden*

Keywords: forest water ecosystems, lime treatment, phytoplankton, macrophytes, zooplankton, benthos, fish

Abstract

Effects on the aquatic biota of lime ($CaCO_3$) application in acidified lakes and streams were studied in a number of waters. After treatment, lime-sensitive species of mosses (*Sphagnum* spp.) decreased, but species such as *Potamogeton natans* and *Myriophyllum alterniflorum* seemed to be favoured. A few years after liming species composition and diversity of phytoplankton, zooplankton and benthic insect larvae were almost identical to that found in oligotrophic and non-acid lakes. Molluscs and benthic crustaceans may have difficulties recolonizing. Reproduction of remaining species of fish was successful as soon as pH increased. High survival of larvae and fry can result in some extremely rich year classes with slow individual growth. In most cases restocking of depleted fish stocks was successful.

Introduction

The acidification of oligotrophic freshwater in North America and Scandinavia must be regarded as one of the most severe environmental problems today. Acidic waters are characterized i.a. by low phosphorus concentrations and elevated levels of metal cations such as Mn, Al, Zn, Cd and Pb (Dickson 1980). Acid lakes also contain reduced number of species at all trophic levels (Almer *et al.* 1978; Hobaek & Raddum 1980; Muniz 1981). These reductions may be caused by direct toxicity of the water (Muniz 1981), decreased leaching of phosphorus from the drainage area (Broberg & Persson 1981), increased competition for phosphate in the lakes (Jansson 1981) or changes in predator–prey relations in the ecosystems (Eriksson *et al.* 1980).

Indications of the effects of acidification on roach *(Rutilus rutilus)* in Sweden (Almer & Hanson 1980) and atlantic salmon *(Salmo salar)* in Norway (Muniz 1981) were already noted in the 1920s. Today, fish stocks are affected in at least four regions in Ontario, Canada (Harvey *et al.* 1981) and in Norway an area of 13 000 km² is devoid of fish populations, and damage is evident over a further 20 000 km² (Sevalrud *et al.* 1980). In Sweden, it has been calculated that about 18 000 lakes, i.e. 20% of the total number of lakes greater than 1 ha have pH <5.5 at some time during the year, and that fish stocks have been damaged in about 9 000 lakes in the south and middle part of the country (Johansson & Nyberg 1981).

Parallel to earlier begun sulfur emission restrictions, the Swedish Government in 1976 initiated a five-year test period with government grants for liming measures to restore acidified lakes and rivers. The practical aim of the experimental activity was to save waters of special value for fishery, nature conservancy or recreational use and at the same time study the ecological effects of lime treatment. Special awareness was given to observing any negative effects of liming.

The results presented here derive mostly from projects carried out within the test period. They should thus be regarded as short-term effects. The same material has been reported in greater detail in

Hydrobiologia 101, 145–164 (1983).

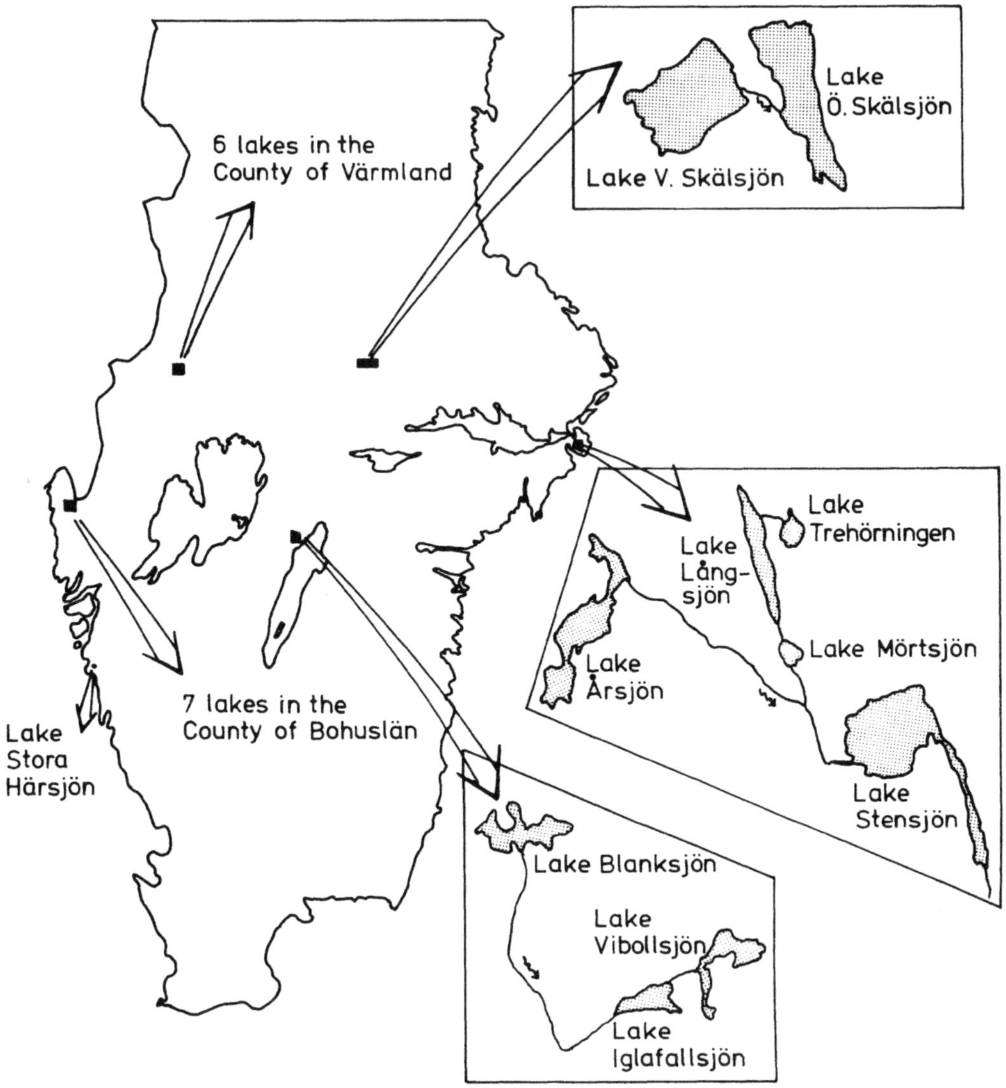

Fig. 1. Location of the lakes studied.

Swedish (Eriksson *et al.* 1982). Another study (Hultberg & Andersson 1981), carried out in four lime-treated Swedish lakes, covers a somewhat longer period, while Scheider & Dillon (1976) give results from one year before and one year after treatment.

Areas of investigation and characteristics of some of the lakes

The location of the lakes investigated is shown in Fig. 1. The lakes are rather small, have low specific conductance, represent clear as well as humic lakes, and are poor in plant nutrients (total nitrogen: 0.19–1.24 mg N · l^{-1}, total phosphorus 2–23 μg P · l^{-1}). pH values before liming were in the range 4.0–5.9 and about 1–2 units higher during the first year after treatment (Table 1). The lowest values were always recorded in surface samples during spring flood, while the highest ones usually were measured during late summer.

Table 1. Size, specific conductance, water colour and pH in some of the lakes studied.

Lake	Surface area (ha)	Spec. cond. ($mS \cdot m^{-1}$)	Water colour ($mgPt \cdot l^{-1}$)	pH before treatment	pH after treatment			
					year 1	year 2	year 3	year 4
Trehörningen	3	4.4–6.8	30–120	4.0–5.3	6.5–6.7	6.1–6.6	5.7–6.2	
Långsjön	10	4.2–5.5	20–65	4.6–5.5	6.3–6.5	6.0–6.5	5.8–6.2	
Mörtsjön*	3	4.3–5.3	15–65	4.9–5.9	6.3–6.5	6.1–6.7	5.9–6.7	
V. Skälsjön	40	2.6–3.3	0–10	4.7–5.5	5.8–6.5	5.4–5.8	5.6–5.9	5.2–6.7
Ö. Skälsjön	34	2.6–3.1	5–10	5.0–5.7	5.7–6.2	5.1–6.0	5.6–6.1	5.2–6.8
St. Sirsjön	25	4.7–6.0	5–20	5.1–5.4	6.5–6.8	6.3–7.0	6.1–7.6	5.8–7.0
Blanksjön	11	3.0–4.2	15–45	4.7–5.4	6.3–6.6	6.0–6.5	5.9–6.3	6.0–6.7
Iglafallssjön	6	2.9–4.1	15–80	4.6–5.0	6.1–6.8	5.2–6.5	5.6–6.3	5.9–6.4
Vibollsjön*	10	2.9–5.5	35–125	4.5–4.9	4.9–5.0	4.9–5.8	4.9–5.5	4.8–5.5
7 lakes in the County of Bohuslän	10–84			5.0–5.2	5.7–7.1	5.8–8.2	5.5–7.0	5.8–6.8
6 lakes in the County of Värmland	25–395			5.0–5.9	6.9–7.2			

* Not limed but influenced by treatment of upstream lakes.

Methods

All lakes studied, except the seven in the County of Bohuslän (Fig. 1), were treated with fine crushed limestone ($CaCO_3$). In the latter, a kind of lime slag ($CaMgSiO_4$) was used. An additional smaller-scale treatment using burnt lime (CaO) was performed in Lake V. Skälsjön in 1976.

Water samples for chemical analyses were taken at three depths in each lake. In the lakes draining to Lake Stensjön (Fig. 1), samples were taken 4–15 times and in the other lakes two times a year, during spring and autumn turnover. In addition, pH and alkalinity were measured once a month in the latter lakes. pH, alkalinity and phosphorus were analyzed according to the methods pH/25, Alk/OS and P-tot/OAP, respectively (SNV 1981).

Phytoplankton samples (50–100 ml), preserved with Lugols' solution, were taken three times in the period July 15–September 15 at a depth of 0.5 m in the middle of the lakes. The samples were analyzed according to Utermöhl.

During late summer the qualitative composition of the planktic community of southern Swedish lakes is rather stable. Analyses of three samples will therefore satisfactorily represent the composition of the phytoplankton community. Moreover, the results in Figs. 2 and 4 are based on samples from a great number of lakes.

Distribution and density of macrophytes was studied along profiles as deeply as was visually possible. At greater depths a Luther-rake was used. Vegetation studies were performed once before and two times after liming.

In lakes Blanksjön, Iglafallssjön and Vibollsjön (Fig. 1), four zooplankton samples, each of 1.6 l, were taken with a Ruttner sampler once a month at depths of 0.5, 2, 4 and 6 m and filtered through a 150 μm mesh. Volume-weighted averages for 5–6 sampling occasions during the period May–October have been calculated. In the other lakes zooplankton was sampled with a net (63–75 μ mesh size) once a year within the period July–September. The occurrence of each species in these samples was analyzed according to a subjective scale where + means less frequent, ++ rather frequent and +++ frequent abundance. All organisms were identified to species except the genus Cyclops and nauplii.

Samples of benthic invertebrates were taken twice a year in April and December in four lakes and once a year in April in five lakes with an Ekman grab. Two samples were taken at a depth of 1 and 2 meters, respectively, and thereafter every other meter to the greatest depth in each lake. The material was then sieved through a 0.6 mm mesh. The relative number of individuals after liming in each lake was calculated in relation to the number of individuals at the corresponding time before treatment. Average values for all nine lakes were calculated.

Test fishing in lakes was usually carried out once a year in July–September using benthic gill nets, 36 m long and consisting of 12 different mesh sizes (10–75 mm) in sections. In streams and rivers, salmonids were sampled by electro-fishing. Fishes were measured from the tip of the nose to the end of the tail. Age was determined from scales *(Salmo trutta, Rutilus rutilus)*, opercular bones *(Perca fluviatilis)* and otoliths *(Salvelinus alpinus)*.

Results and discussion

Phytoplankton

In strongly acid lakes in southern Sweden (pH <5), only 5–20 species of phytoplankton are generally observed in summer samples, in contrast to 30–80 in lakes with pH >6 (Fig. 2). Investigation of these acid lakes after lime treatment has shown that the number of species has increased and the lakes have gradually obtained a taxonomic phytoplankton composition similar to that in adjacent unlimed oligotrophic lakes, at about pH 6. The acid-tolerant and extremely oligotrophic species comprising the bulk of biomass in the acid lakes, especially *Gymnodinium uberrimum* (Allm.) Kof & Swezy and *Peridinium inconspicuum* Lemm. (Dinophyceae) have been replaced primarily by chrysophyceans,

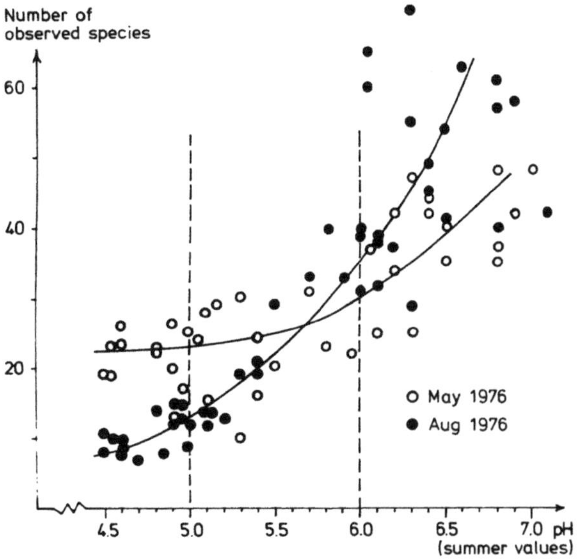

Fig. 2. Number of observed species of phytoplankton in May and August 1976 in relation to pH in Swedish west coast lakes.

Table 2. Concentration of total phosphorus and percentage of acid tolerant dinophyceans *(Gymnodinium uberriumum* and *Peridinium inconspicuum)* of the total phytoplankton biomass inlakes with pH >6.

Total P $\mu g \cdot l^{-1}$	≤5	6–7	8–12	>12
Acid tolerant dinophyceans (%)	34	17	2	2
Total biomass ($\mu g \cdot l^{-1}$)	233	279	584	1 424
No. of lakes	15	22	15	16

but also by diatoms and Chlorophyceae.

The described dinophycean dominance in the acid lakes is apparently mainly a result of nutrient depletion since this group is less common when the nutrient supply increases. This has also been confirmed by investigations in more alkaline lakes (pH >6), where dinophyceans occurred frequently at low phosphorus concentrations (Table 2).

In Sweden, the period 1975–76 was characterized by very low precipitation and the pH values of the lakes were relatively high. In 1977, however, there was a considerable increase of precipitation, but leaching of phosphorus was low, probably because of high concentrations of reactive aluminium in the soil. The following years (1978–80) received similar amounts of precipitation. pH of inflowing water, however, gradually rose and so did leaching of phosphorus. This raised the algal production considerably. Thus phosphorus concentration and pH increased also in unlimed lakes, here illustrated (Table 3) by values from Lake Årsjön, a small and virgin forest lake (Fig. 1). In most lakes in central Sweden algal production increased in 1979, but in lakes in the west coast area not until 1980.

In view of these observations it is not surprising that the character of the phytoplankton remained acidic in most limed lakes during the first summer after liming (1978), since extremely poor nutrient conditions prevailed. A case in point are the lakes Blanksjön and Iglafallssjön which were limed during spring 1978. The treatment raised the pH value the following summer from 4.9 to about 6.7 while

Table 3. pH and concentration of total phosphorus ($\mu g \cdot l^{-1}$) in Lake Årsjön (summer values).

Year	1974	1975	1976	1977	1978	1979	1980
pH	5.5	5.4	6.0	5.3	5.7	6.0	5.4
Total P	6	7	–	4	5	11	8

Table 4. pH, total phosphorus ($\mu g \cdot l^{-1}$), trophic index and the observed number of phytoplankton species in summer samples from lakes Blanksjön, Iglafallssjön and Vibollsjön.

Lake		1977	1978	1979	1980
Blanksjön	pH	4.9	6.6	6.3	6.3
	Total P	3	2	6	11
	No. of species	27	25	47	–
	Index	20	19	28	–
Iglafallssjön	pH	4.9	6.8	6.1	6.3
	Total P	5	4	11	8
	No. of species	29	26	42	45
	Index	20	16	25	25
Vibollsjön	pH	4.8	4.9	5.1	5.3
	Total PO	7	4	12	–
	No. of species	22	25	50	–
	Index	23	23	29	–

Lake Vibollsjön, situated below and receiving some water from these lakes, had a pH rise from 4.8 in 1977 to 5.1 in 1979 (Table 4).

The results of liming these acid lakes give an idea of the relative importance of nutrient supply and pH value.

During summer 1978 there was no effect on phytoplankton in Blanksjön and Iglafallssjön in spite of the raised alkalinity.

When the phosphorus supply increased in 1979, however, there was an extensive qualitative change in phytoplankton composition, which became dominated by diatoms and chrysomonads. In acid Lake Vibollsjön, however, the biomass was dominated by the acid-tolerant flagellate *Gonyostomum semen* (Ehr.) Dies. (Rosén 1981). Most remarkable was the observation of 50 species at such a low pH, a unique occurrence during this study (Table 4).

In most lakes studied the lime treatment had no or very little effect upon the phytoplankton until the phosphorus supply was markedly raised. Becuase of this, the results from only a few lakes can be used to determine whether a pH rise itself, irrespective of the nutrient level, will change the phytoplankton composition of the acid lakes. These waters are shallow with a short retention time. Such an example is the north basin of Lake Stora Härsjön in which a very marked qualitative change occurred shortly after the lime treatment. The new phytoplankton during the summer of 1978 had a most abnormal composition and few species (Fig. 3), a phenomenon also observed in other lakes of this quickly reactive type. It is worth noting that the new

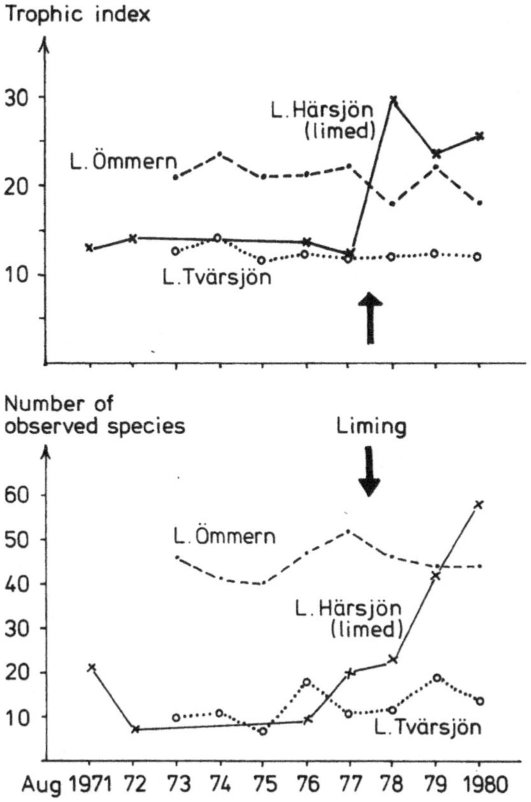

Fig. 3. Trophic index and number of observed species of phytoplankton in lakes St. Härsjön (limed), Ömmern and Tvärsjön (untreated) in August 1971–1980.

species, both during this summer and in 1979 when the plankton composition became normal, were by experience known to have a higher nutrient demand and a faster generation turnover than the acid dinophyceans. This would indicate that there had been an increased mineralization and recycling of nutrients and thus a raised primary production.

The results suggest that the rapidity of the change in phytoplankton composition is mainly dependent on the mixing of the water. In agreement with this was the observation that the number of species in acid lakes is less divergent from those having a normal pH in spring (Fig. 2) when there is melting of snow and temperature induced vertical water movements.

In investigating the changing nutrient supply, or the trophic level of these lakes, the qualitative information given by the phytoplankton species was most valuable, since most species have a relatively narrow trophic niche. The use of a trophic index based on the occurrence and frequency of phytoplankton indicator species (Hörnström 1981) has, especially in the case of Lake Stora Härsjön (Fig. 3), given evidence that liming has increased the primary production of phytoplankton during the late summer period. The trophic index in this lake increased from 12 to about 25 and the number of species from 20 to 50. During the second summer after the liming the plankton composition was very similar to that in adjacent pH-normal lakes.

The first phytoplankton species to appear after lime treatment in the strongly acid lakes were species of which a few cells may also accidentally occur in unlimed acid waters when there is a raised nutrient supply. On the other hand, it is notable that many, markedly oligotrophic species, common in adjacent lakes were absent in the acid lakes both before and after liming. These species were probably completely eliminated during the acid period and must return gradually, by airborne means.

Thus, even if the phytoplankton composition is mainly a result of the nutrient supply, this is obviously not the whole explanation. pH in itself is crucial. It is well known, for instance, that several cyanophyceans require relatively alkaline waters. Another fact likely to be of considerable importance is the high level of toxic models (which is lowered as a result of liming). Laboratory tests were performed on an isolated oligotrophic green algae, absent from the acid lakes. Its growth was completely inhibited by an aluminium concentration of 250 $\mu g \cdot l^{-1}$, corresponding to that of acid lakes, but the alga grew well in lake water (pH 5.8) low in aluminium with pH adjusted to 5.0 (Hörnström *et al.*, in prep.).

The additional combined stress, exerted by low pH, nutrient depletion and metal toxicity may not immediately cease after liming since such treatment is generally just confined to the lakes. As the drainage area is usually not treated phosphorus will still be bound to the surrounding soil and the metal supply to the lake will continue. Thus there may be several reasons for the relatively low number of species found in most acid lakes during the first years after liming (Fig. 4).

Macrophytes

The vegetation before liming of the lakes studied was such as will usually be found in oligotrophic lakes with a rather low pH (Iversen 1929; Lohammar 1938, 1965). Lake Långsjön had as many as 22 species while Lake Trehörningen, the most humic lake, had only 11 (Table 5).

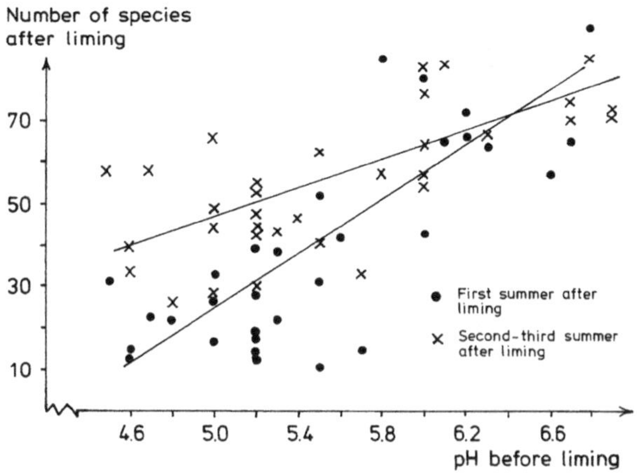

Fig. 4. Number of species of phytoplankton after liming in relation to pH before liming.

In all the lakes except St. Sirsjön the bottom was covered by vegetation to more than 40% and in Lake Långsjön it was covered by nearly 70%. The isoetids and the mosses were the most dominating plants (Table 6). Mats of isoetids covered large areas in all the lakes except in Lake Trehörningen, where they were absent probably due to the loose muddy bottom of this lake. Compact blankets of mosses, especially *Sphagnum* and *Fontinalis*, were found in all the lakes, and they grew on more than 50% of the area covered by vegetation in lakes Trehörningen, Långsjön and V. Skälsjön. The occurrence of *Sphagnum* in acidified lakes is well known, and the abundance of *Sphagnum* seems to be negatively correlated to the pH of the lake water (Grahn *et al.* 1974; Grahn 1977). Helophytes, nymphaeids and elodeids were distributed mostly in small areas of the lakes studied (Table 6).

The helophytic vegetation has changed to only a small extent after the lime applications. Common species such as *Carex rostrata, Equisetum fluviatile* and *Phragmites australis* have the same areas of distribution as earlier, but the stands of *Carex* spp. seem to be more dense and the shoots are usually bigger in the limed parts of the inner littoral zone. The latter condition indicates that an increased recycling of nutrients has taken place in this part of the littoral.

In most of the lakes the nymphaeids have about

Table 5. List of macrophyte species in the lakes treated with lime.

	Trehörningen	Långsjön	St. Sirsjön	V. Skälsjon
Helophytes				
Carex rostrata	0	0	+	0
Carex lasiocarpa	0	0	+	0
Carex vesicaria		0		
Equisetum fluviatile	0	0	0	0
Phragmites australis	0	0		0
Eliocharis palustris		0		
Menyanthes trifoliata	0		0	0
Typha latifolia		0		
Lythrum salicaria		0		
Lysimachia thyrsiflora		0	0	0
Nymphaeids				
Glyceria fluitans		0		
Potamogeton natans	*	*	0	
Nymphaea alba	+	0	0	0
Nuphar lutea	+	0	0	0
Sparganium spp	+	+	0	0
Elodeids				
Myriophyllum alterniflorum		+	0	
Juncus bulbosus		0	–	0
Utricularia vulgaris	+		0	
Isoetids				
Lobelia dortmanna		0	0	0
Isoetes lacustris		0	0	0
Littorella uniflora		0	0	0
Ranunculus flammula		0		
Mosses				
Drepanocladus trichophyllus		0		+
Drepanocladus sp	0		0	0
Fontinalis antipyretica		0		0
Fontinalis dalecarlica		0		
Sphagnum subsecundum (coll.)	–	–	–	–

0 = Unchanged distribution area and density.
+ = Increased distribution area and/or density after liming.
– = Decreased distribution area and/or density after liming.
* = New species after liming.

Table 6. Distribution areas (ha) of macrophytes before and after the lakes had been treated with lime.

	Trehörningen		Långsjön		St. Sirsjön		V. Skälsjön	
	1976	1980	1976	1980	1976	1980	1976	1980
Lake area	3.4		10.2		24		40	
Vegetation area	1.7	1.4	6.9	6.9	5.4	5.1	17.4	9.3
Helophytes	0.4	0.4	1.2	1.2	0.8	0.8	1.6	1.6
Nymphaeids	0.5	0.8	1.3	1.3	1.0	1.0	0.5	0.5
Elodeids	0	0	0.1	0.2	0.7	0.5	0	0
Isoetids	0	0	3.7	3.7	2.5	2.5	4.2	4.2
Mosses	0.8	0.3	3.1	3.1*	1.2	0.8	11.9	4.0
Sphagnum	0.8	0.3	2.0	2.0*	1.2	0.8	11.9	2.6

* Mostly solitary thin stands of *Sphagnum*.

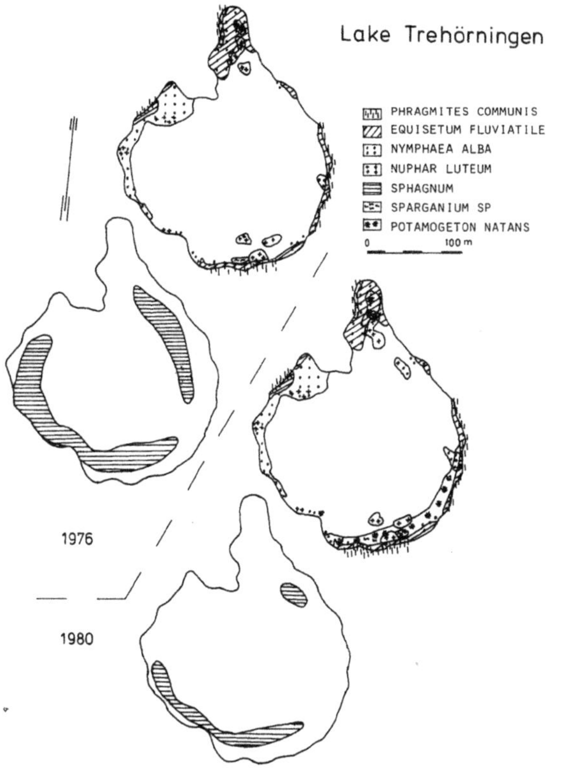

Fig. 5. Distribution areas of macrophytes in Lake Trehörningen before and two years after liming.

the same distribution areas as earlier (Table 6). Small increase in *Nuphar lutea, Nymphaea alba* and *Sparganium* spp. did occur in Lake Trehörningen and Lake Långsjön, however, these changes probably lay within the range of natural variation. For example, the occurrence of *Sparganium angustifolium* can vary greatly within a period of a few years (Eriksson *et al.* 1974). *Potamogeton natans* did appear after liming in the same lakes, and it has also increased its distribution area considerably in Lake Trehörningen during the last vegetative period (Fig. 5). The plant might have been present in the lakes at an earlier time, or, it may have been spread to the lakes by birds from a neighbouring lake, where it is very abundant. Since its seeds can germinate after several years storage in the sediment (Hutchinson 1975), liming might have changed the conditions favoring the germination of the seeds. Although *P. natans* is considered to be eurytopic with respect to pH by several authers (see Hutchinson 1975) it seems to be absent from very acidic waters (Iversen 1929; Sand-Jensen & Rasmussen 1978), thus it could be favored by the increased pH in Lake Trehörningen.

After the liming, the submerged form of *Juncus bulbosus* decreased considerably in Lake St. Sirsjön, where pH increased to 7. Since submerged forms of amphibious plants seem to have a high demand of free CO_2 (Bristow 1969), the increase in pH is likely to be the reason for the decreased abundance of *J. bulbosus* in Lake St. Sirsjön. The submerged form of *J. bulbosus* is very common in acidic lakes (Hinneri 1976; Almer *et al.* 1978) and it is one of the few macrophytes inhabiting strongly acidic streams (Sand-Jensen & Rasmussen 1978).

Myriophyllum alterniflorum appeared as a few individuals in Lake Långsjön. It increased directly after liming, and the stands were big and dense even two years after treatment. Its natural variation is rather great (Macan 1977), but it is generally uncommon in acid lakes (Iversen 1929). It usually does not grow in big dense stands in lakes with a pH below 5. The increase could be an effect of the rise in pH.

The isoetids do not seem to have been affected by liming. Common species such as *Isoetes lacustris, Lobelia dortmanna* and *Littorella uniflora* occupy the same distribution areas as before (Table 6), and, so far, they have probably been favored by the increase in pH. An increased mineralization of the organic matter that accumulates in the littoral zone of acidified lakes and a decreased competition from *Sphagnum,* the latter seeming to be sensitive to liming, must be advantageous for this type of vegetation.

After lime treatment, *Sphagnum* decreased considerably in all the lakes, in the deeper parts as well as in the littoral zones. The deeper parts of Lake V. Skälsjön were limed with unslaked lime in 1976. A large part of it probably ended up on the bottom and it must have had an immense effect on the mats of *Sphagnum* which disappeared completely (Fig. 6). Treatment of the littoral zones with fine crushed limestone also produced negative effects on *Sphagnum,* which was rapidly killed in those areas more or less covered with lime.

Merely an increase in pH also seems to have a negative effect on *Sphagnum.* Growth was very stunted compared to that under acid conditions. The stands became thin and scanty, and in Lake Trehörningen they partly disappeared from their earlier area of distribution (Fig. 5).

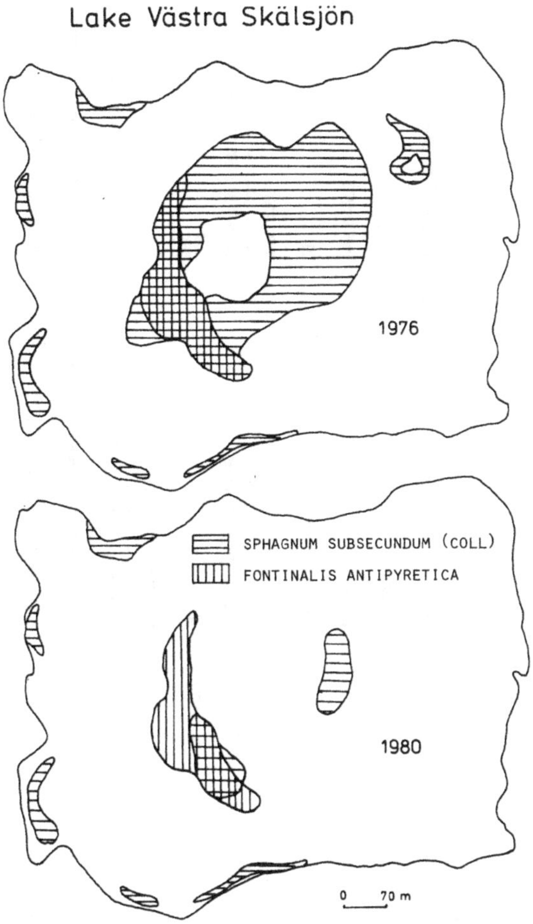

Fig. 6. Distribution areas of *Sphagnum subsecundum* (coll.) and *Fontinalis antipyretica* before and four years after lime treatment in Lake V. Skälsjön. In 1978 *S. subsecundum* (coll.) was totally absent from the central part.

Sphagnum has started to recolonize the deeper parts of Lake V. Skälsjön, where pH has stayed about 6. Two very thin stands were observed in 1980 (Fig. 6).

Other species of mosses, like *Fontinalis* spp. and *Drepanocladus* spp., have not indicated any sensitivity to liming. They maintained the same areas of distribution and their stands were about as dense as earlier. *Drepanocladus trichophyllus* was probably spread over rather large areas in Lake V. Skälsjön before *Sphagnum* invaded.the deeper parts of the lake. Before liming, dead stands of *Drepanocladus trichophyllus* were found under the mats of *Sphagnum* and living plants were found very sparsely intermixed with *Sphagnum*. In 1980 *Drepanocladus trichophyllus* grew sparsely over the whole area earlier dominated by *Sphagnum* and it must have been favored by the decreased competition from *Sphagnum*. Dead stands of *Scorpidium scorpioides* were also found under the mats of *Sphagnum*. This moss species is not uncommon in lakes with a pH above 6 (Iversen 1929; Sörensen 1948) and it may have grown in rather extensive stands in the fourties when the lake, according to Brundin (1949), had a pH above 6. *Scorpidium scorpioides* has been found to decrease considerably in several lakes when the pH has decreased below 6 (Eriksson, unpubl.).

Zooplankton

Effects of lime application on the rotifer community have been studied in 7 lakes in the County of Bohuslän and 6 lakes in the County of Värmland (Ekström, unpubl.). Dominating species in the Bohuslän lakes during acid conditions were *Keratella quadrata, K. cochlearis, Kellicottia longispina* and *Polyarthra remata* (Fig. 7a), but the abundance

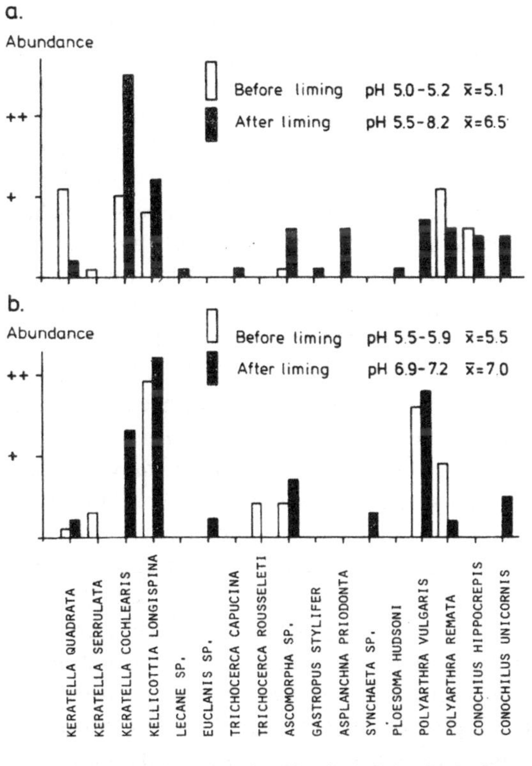

Fig. 7. Abundance of species of rotifers in the lakes in the County of Bohuslän (a) and the County of Värmland (b) before and after lime treatment.

of even these species has been regarded as less frequent. *Kellicottia longispina* and *Polyarthra vulgaris* varied from less to rather abundant and were dominating species under similar conditions in the lakes of the County of Värmland (Fig. 7b). Most species found and especially the dominant ones seem to be common in oligotrophic lakes and may be regarded as eurytopic (Pejler 1957).

After liming the planktic rotifer fauna was more or less dominated by the same species as before treatment in all the waters, and the most marked effect was an increased abundance of above all *Keratella cochlearis* and *Kellicottia longispina*. Henrikson *et al.* (1981) found a massive development of *K. cochlearis* and *Polyarthra* spp. in Lake Stensjön after lime application. In this lake (op. cit.) and in the four lakes studied by Hultberg & Andersson (1981) and the lakes of the County of Bohuslän a marked increase of *Asplanchna priodonta* was also noted.

As sampling intensity was somewhat unsatisfactory, the conclusions presented here are only tentative. However, the main reason for the increasing numbers of individuals of rotifers may be an elevated supply of suitable food after lime treatment. *Keratella cochlearis* and *Conochilus unicornis-hippocrepis* are filtering species living on detritus and bacteria (Pourroit 1977). Increasing amounts of heterotrophic bacteria after liming has been found by Scheider & Dillon (1976), while Traaen (1980) found no differences in numbers of pelagic bacteria in acid and non-acid lakes. *Kellicottia longispina* and *Polyarthra* spp. feed mainly on small species of phytoplankton (Pourriot 1977) and the changing composition of the phytoplankton community is probably the reason for the development of these species. The great abundance of especially *Keratella cochlearis* and *Polyarthra* after liming ought to favour *Asplanchna priodonta*, which feeds on i.a. those species (Guiset 1977; Pourriot 1977).

The occurrence of observed species of planktic crustaceans before and after lime treatment in the earlier discussed lakes in the Counties of Bohuslän and Värmland is shown in Fig. 8 (Ekström, unpubl.). Copepods and *Bosmina coregoni* dominated the planktic fauna before liming. Even if one sampling occasion per year does not permit any extensive conclusions, the results indicate that the abundance of most observed species increased after treatment. This was particularly obvious for *Diaphanosoma brychyurum* and nauplii in Bohuslän (Fig. 8a) and *Daphnia* spp. and *Cyclops* spp. in Värmland (Fig. 8b). Similar results, i.e. increasing number of species of cladocerans and biomasses of planktic crustaceans, were found by Hultberg & Andersson (1981). In another study (Scheider & Dillon 1976) the biomass decreased, probably due to a too rapid pH rise, caused by the reactive agent used ($Ca(OH)_2$).

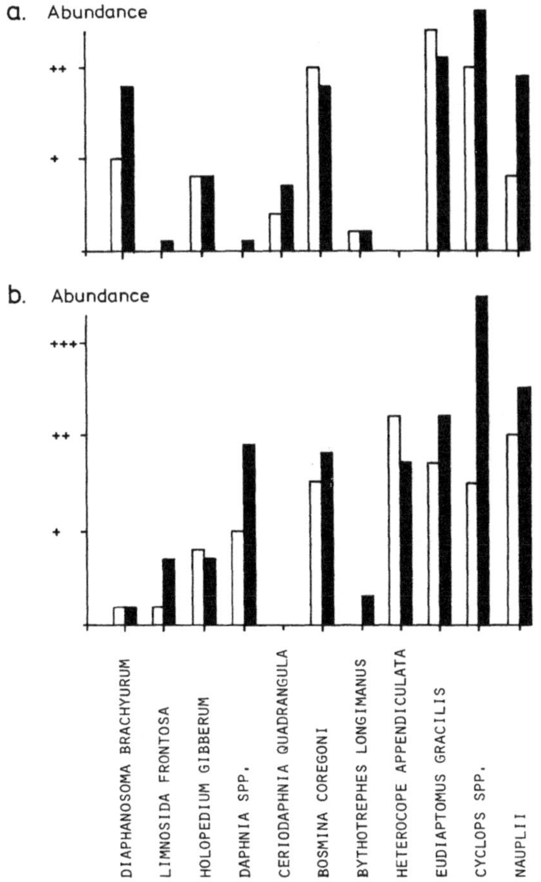

Fig. 8. Abundance of species of planktic crustaceans in the lakes in the County of Bohuslän (a) and the County of Värmland (b) before and after lime treatment (symbols as in Fig. 7).

The composition of the planktic fauna of crustaceans in acidic lakes is probably to a great extent an effect of selective predation by invertebrate predators such as *Chaoborus* spp. and *Glaenocorisa p. propinqua* (Eriksson *et al.* 1980). The same predators could obviously also prevent the development of most planktic cladocerans after lime treatment. The acidic lakes Blanksjön, Iglafallssjön and Vibollsjön (Fig. 1) had been barren of fish for at least

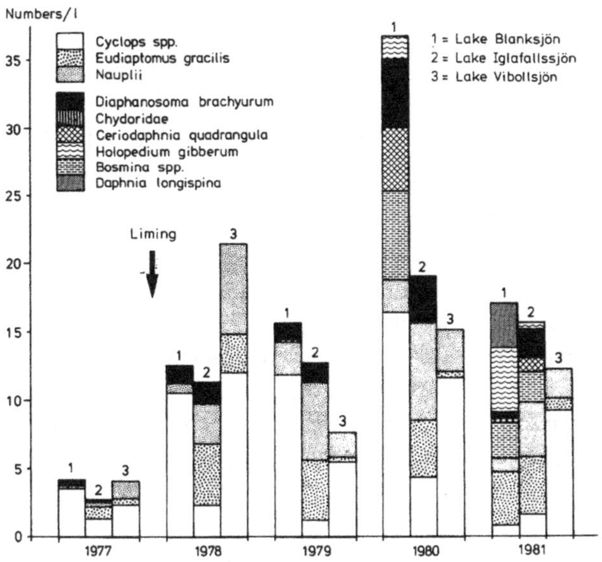

Fig. 9. Average number of planktic crustaceans in May–October in lakes Blanksjön, Iglafallssjön and Vibollsjön.

ten years and in all of them great abundances of *Chaoborus flavicans* and smaller densities of *C. obscuripes* were found. The two first mentioned lakes were treated with lime in 1978. Lake Vibollsjön has not been treated, but is affected by limed runoff water from the two upstream lakes. The average number of planktic crustaceans during May–October increased manyfold after lime application (Fig. 9). Even in 1979, however, the numbers of cladocerans were still small in all three lakes. In that summer fry and yearlings of brown trout *(Salmo trutta)* were stocked in Lake Iglafallssjön and Blanksjön, respectively. The yearlings in Lake Blanksjön intensively preyed upon and rapidly eliminated the chaoborid population and in 1980 and especially in 1981 the crustacean community was dominated by cladocerans (Fig. 9). The chaoborid population in Lake Iglafallssjön was eliminated first in 1980 and in 1981 several species of cladocerans began to appear. Lake Vibollsjön is still barren of fish, has a dense population of chaoborids and a planktic community consisting only of copepods (Fig. 9), in spite of a better water quality and a changing phytoplankton community.

The earlier mentioned changes in the species composition of phytoplankton favors filtering species like cladocerans and ought to be the main reason for their increase after lime treatment. The slow increase of cladocerans found in some lakes probably depended on intensive selective predation from rich year classes of different species of fish or dense population of chaoborids or other predatory invertebrates. In most lakes, however, both filtering rotifers and crustaceans increased, indicating improved production of bacteria and edible algae after lime treatment.

Benthos

The benthic community is very complex and can be affected by acidification and liming in several ways. Many benthic invertebrates like molluscs (J. Okland 1980; K. A. Okland 1980) and certain crustaceans (K. A. Okland 1980) and ephemeropterans (Johansson & Nyberg 1981) are highly susceptible to a low pH value. Other changes in the community may be indirect and caused by altered predator–prey relations (Eriksson *et al.* 1980).

Asellus aquaticus seem to be a very pH-tolerant crustacean (Mossberg & Nyberg 1979; K. A. Okland 1980), but the species was apparently greatly favored by liming. The first autumn after treatment the number of *A. aquaticus* was on the average about 3 times as high as during the corresponding time before treatment (Fig. 10). The following spring, however, the abundance was much lower than during pre-liming conditions. The difference in abundance between the following sampling occasions was somewhat less striking and the population tended to stabilize at a slightly higher density compared to pre-liming conditions. Increasing abundances of benthic crustaceans after lime treatment have also been reported by Hultberg & Andersson (1981) *(Asellus aquaticus* and *Pallasea quadrispinosa)* and Fiskeristyrelsen and Statens Naturvårdsverk (1981) *(Astacus astacus).*

The only species of Ephemeroptera found in the

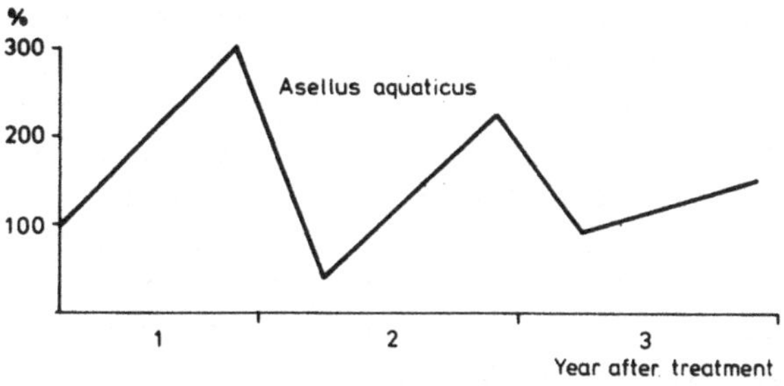

Fig. 10. Development of *Asellus aquaticus* in the littoral zone after liming (in percent of pre-liming values; average values for 9 lakes).

lakes studied before liming was the acid-tolerant *Leptophlebia vespertina* (Hendrey & Wright 1976; Mossberg & Nyberg 1979). The species generally became more abundant (+142%) after treatment. *Cloeon dipterum* appeared three years after liming in four of the lakes, probably due to an increased pH. The same species was also found by Hultberg & Andersson (1981) after treatment.

Members of the order Oligochaeta were not identified to species in the present study. Regarded as a group they increased in numbers both in littoral and profundal samples in most lakes and on the average the number approximately doubled three years after liming. In the profundal of the two most oligohumic lakes (lakes V. and Ö. Skälsjön) the number of oligochaetes decreased by about 40% after treatment.

Species of the above-mentioned orders seem to feed on detritus (*A. aquaticus* and *L. vespertina*; Kjellberg 1972; Moore 1975) and bacteria (Oligochaeta; Brinkhurst & Jamieson 1971). Traaen (1980) found that high pH-enhanced bacterial decomposition of glucose and glutamic acid and Gahnström *et al.* (1980) found higher oxygen consumption as well as a more rapid turnover rate of glucose in the sediment after lime treatment. The probable improved supply of bacteria in detritus and sediment after liming ought to be the primary reason for the increasing population of oligochaetes. The rapid increase of *A. aquaticus* and even of *L. vespertina* after treatment might be suspected to depend on reduced mortality. This might have been caused by reduced predation, but as the abundance of invertebrate predators was unaltered or slightly raised after treatment (Fig. 12) and predation by fish was increased due to successful recruitment or restocking, this does not seem to be the reason. Even if *A. aquaticus* and *L. vespertina* are found in acidic lakes, recruitment may be reduced at low pH. Survival is also favored by improved food conditions, perhaps in the form of bacteria and other organisms attached to detritus. The marked increase of these species and even of Oligochaeta during the first summer after treatment, certainly is the effect of a changed exploitment of organic material accumulated during acid conditions (Grahn *et al.* 1974). The drastic decrease of the *Asellus* population during the first winter after liming was probably due to a shortage of food caused by high population density. Increased numbers of the organisms mentioned, in spite of increased predation on *A. aquaticus* and *L. vespertina*, indicate increased production after lime treatment.

Chaoborus flavicans increased considerably after liming both in benthic (Fig. 11) and planktic samples (Nyberg, unpubl.). Chaoboridae are feeding on smaller zooplankton (Fedorenko 1975), and the increasing abundance of these organisms certainly was the reason for the development of *Chaoborus flavicans*. The most marked increase was noted in the lakes barren of fish. In the other lakes restocking of fish and successful recruitment after liming resulted in reduced number of *C. flavicans* (Fig. 11) since the species is very susceptible to fish

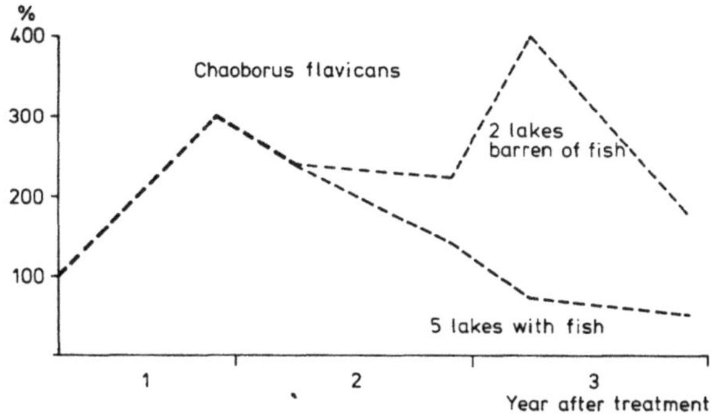

Fig. 11. Development of *Chaoborus flavicans* in the profundal zone after lime treatment (in percent of pre-liming values; average values for 9 lakes).

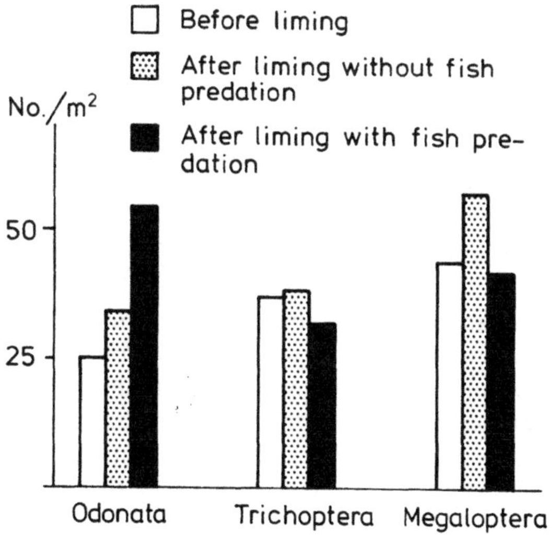

Fig. 12. Abundance of Odonata, Trichoptera and Megaloptera (*Sialis lutaria*) before and after treatment (average values for 9 lakes).

Table 7. Number of lakes where certain species of Trichoptera were found before (BL) and after (AL) liming. A total of 18 lakes (the nine lakes of this investigation, Lake Ölen (Bergquist 1980) and eight lakes in the County of Bohuslän (Ekström, unpubl.)).

Species	BL	AL
Agrypnia varia	6	6
A. obsoleta	3	4
Cyrnus flavidus	6	5
Holocentropus dubius	9	9
Limnophilus sp.	4	6
Molanna angustata	1	5
Mystacides azurea	2	2
Nemotaulius punctatolineatus	1	2
Phryganea bipunctata	1	1
Plectrocnemia conspersa	1	0
Polycentropus flavomaculatus	3	4

predation (Eriksson *et al.* 1980).

Species of the orders Odonata, Trichoptera and Megaloptera *(Sialis lutaria)* are often regarded as susceptible to fish predation (op. cit.) and thus favored by acidification (Grahn *et al.* 1974). The number of Trichoptera was practically unaffected by liming (Fig. 12) and even the species composition changed very little after lime treatment. The only species that showed any marked positive reaction was *Molanna angustata,* which was found only in one lake before and in five lakes after treatment (Table 7). Odonats increased after liming both in lakes barren of fish and, curiously enough most in lakes with fish predation (Fig. 12). This surprising result can either be explained by the fact that the first mentioned lakes only were barren during 1–2 years after treatment, perhaps too short a period for complete recovery, or by differences in productivity between the two groups of treated lakes. Anyhow, fish predation has not had any serious effect on the number of odonats in these lakes so far. *Sialis lutaria* increased after liming in the lakes barren of fish and was about the same as before treatment in the other lakes. Unaltered or increased numbers of individuals of these orders after liming, in spite of an increased predation by fish, indicate an increased production, probably due to an elevated supply of food.

Before lime treatment Gastropoda were totally absent in the lakes studied and Lamellibranchia were only represented by the genus *Pisidium.* This genus increased after treatment in all lakes except in two of the most humic ones (Table 8). The gastrop-

Table 8. Number of *Pisidium* sp. (ind. · m^{-2}) before (BL) and after (AL) lime treatment.

Lake	Littoral		Profundal	
	BL	AL	BL	AL
V. Skälsjön	0	3	8	52
Ö. Skälsjön	45	73	24	97
St. Sirsjön	102	338	49	159
Blanksjön	0	23	0	0
Mörtsjön	261	201	0	0
Långsjön	7	55	7	8
Iglafallssjön	10	0	0	0
Trehörningen	13	54	0	0
Vibollsjön	0	0	0	0

od *Lymnaea peregra* was found in one of the lakes three years after liming. Mollusca ought to be directly favored by rising pH value and calcium content of the water, but dispersal problems may be of great importance for the recolonization of this order.

Chironomidae consists of many species with often very specific demands on the environment and many of these species also react strongly to acidification and liming.

Tanypodinae, however, showed no general response in all the lakes. The species generally decreased in the littoral and increased in the profundal zone. Dominating genera were *Ablabesmyia* and *Procladius* and they seemed to respond similarly.

The relative part of some species of Orthocladiinae in relation to the total benthic chironomid fauna in the littoral is given in Table 9. After liming *Heterotanytarsus apicalis* became much more abundant in the most oligohumic lakes and in two of the most humic lakes the species was only detected after liming. *H. apicalis* has earlier been found to be disfavored by low pH, and especially by low pH and high humic content (Mossberg 1979; Raddum & Saether 1980). The development of *Heterotrissocladius marcidus* after lime treatment was mostly positive. In Lake V. Skälsjön two other species of this genus, *H. maeri* and *H. subpilosus,* were found after liming. Both, but particularly *H. subpilosus,* are indicators of oligotrophic conditions (Saether 1979). The genus *Psectrocladius* is common in acid and humic waters (Brundin 1949; Mossberg 1979). After liming the genus decreased in all lakes except in the two most humic ones. *Zalutschia zalutschico-*

Table 9. The relative part (in percent) of some species of Chironomidae in relation to total number of sediment living chironomids in the littoral.

Lake	*Heterotanytarsus apicalis*		*Heterotrissocladius marcidus*		*Psectrocladius* spp.		*Tanytarsus* spp.	
	BL	AL	BL	AL	BL	AL	BL	AL
Västra Skälsjön	<1	21	9	17	15	4	6	8
Östra Skälsjön	5	11	0	3	2	0	48	13
Stora Sirsjön	<1	<1	<1	2	8	1	2	39
Blanksjön	0	0	0	0	12	0	2	3
Mörtsjön	1	<1	0	<1	1	1	34	30
Långsjön	4	3	3	6	24	7	<1	24
Iglafallssjön	0	1	3	3	2	1	<1	17
Trehörningen	0	1	<1	<1	4	7	0	36
Vibollsjön	0	0	<1	0	<1	3	0	0

Table 10. The proportion (in percent) of *Phaenopsectra* sp. to the sum of *Chironomus* sp. and *Phaenopsectra* sp. before (BL) and after (AL) lime treatment.

Lake	BL	AL
V. Skälsjön	100	100
Ö. Skälsjön	100	100
St. Sirsjön	6	25
Blanksjön	89	90
Mörtsjön	0	0
Långsjön	0	8
Iglafallsjön	93	97
Trehörningen	0	1
Vibollsjön	4	0

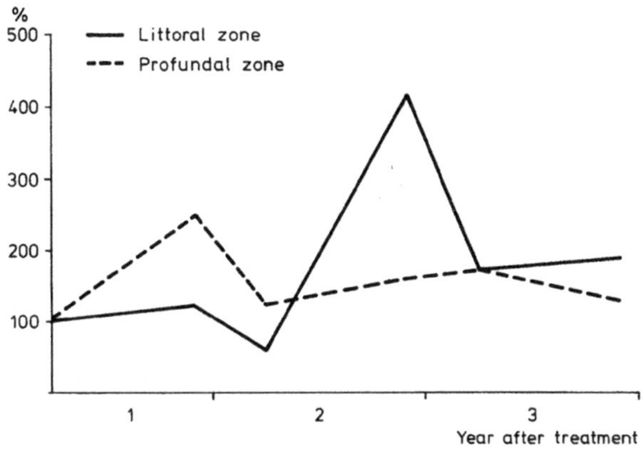

Fig. 13. Development of the benthic community after liming (in percent of pre-liming numbers; average values for 9 lakes).

la and *Z. mucronata* occurred together in lakes Långsjön and Mörtsjön. After treatment *Z. zalutschicola* increased and *Z. mucronata* decreased, which together with the decreased abundance of *Psectrocladius* sp., is a clear indication of a fauna less affected by humic substances (Saether 1979).

The profundal fauna in acid lakes is often dominated by species of the genus *Chironomus* (Mossberg & Nyberg 1979) and in lakes where it occurred, *Chironomus* decreased after liming, but the genus was still abundant. Average values before and after liming were 53 and 44% of total sediment living chironomids respectively. Even *Phaenopsectra,* which is replaced by *Chironomus* at low pH (op. cit.), changed very little after liming. However, in relation to *Chironomus, Phaenopsectra* sp. seemed to increase after treatment (Table 10). The same was valid for *Stictochironomus rosenschöldi* in relation to *Phaenopsectra* sp. in the two most oligohumic lakes. *S. rosenschöldi* is a species indicating oligotrophic conditions (Saether 1979).

The importance of Tanytarsini in the littoral increased in most of the lakes after treatment, especially in Lake Trehörningen and Lake St. Sirsjön (Table 9). The group is known to be sensitive to acidification and humic water (Brundin 1949; Mossberg 1979).

In the littoral zone liming usually led to higher number of benthic invertebrates, with the exception of the first spring (Fig. 13).

The total number of benthos in the profundal zone generally increased during the first period after liming (Fig. 13). This was to a great extent caused by an increasing population of *Chaoborus flavicans.* In the following the average values vary between 127–175% of pre-liming values. The lowest values were found in lakes V. and Ö. Skälsjön, while the most positive response was found in Lake

Trehörningen. Before liming benthic chironomids were almost completely absent from the profundal zone in this lake. After treatment there was a succession of *Chironomus plumosus* gr., *C. anthracinus* gr. and *Zalutschia zalutschicola*.

In the profundal zone of the most oligohumic lakes there is some evidence for a slightly more oligotrophic fauna after liming. This is evidenced by the lower numbers of benthos and occurrence or an increase in the numbers of species of chironomids indicating oligotrophic conditions. One explanation for this might be an increasing decomposition in the epilimnion (Scheider & Dillon 1976) resulting in a reduced sedimentation of organic material to the bottom.

According to Wiederholm & Eriksson (1977) and Mossberg (1979) acidification leads to a benthic community like that found in humic lakes. The results of this investigation indicate that lime treatment led to the formation of a community affected to a lesser degree by humic substances. An example of this is the development of *Psectrocladius, Zalutschia zalutschicola, Z. mucronata* and *Tanytarsus*.

Fish

Recruitment failure due to mortality of eggs and larvae is considered to be the main reason for the extinction of fish populations in acidic water (Muniz & Leivestad 1980). Already in the 1920s lime was used to improve survival of eggs and larvae of atlantic salmon *(Salmo salar)* in Norwegian hatcheries (Leivestad *et al.* 1976).

Reproduction seems to be successful in all waters as soon as pH rises after lime treatment. The spawning of roach *(Rutilus rutilus)* in Lake Mörtsjön had not resulted in any offspring since 1969 and in 1978 the population was made up of few and very old individuals (Fig. 14). After liming the same spring survival of the new year class was very good and in 1981 the gill-net catch was dominated by individuals hatched after the treatment.

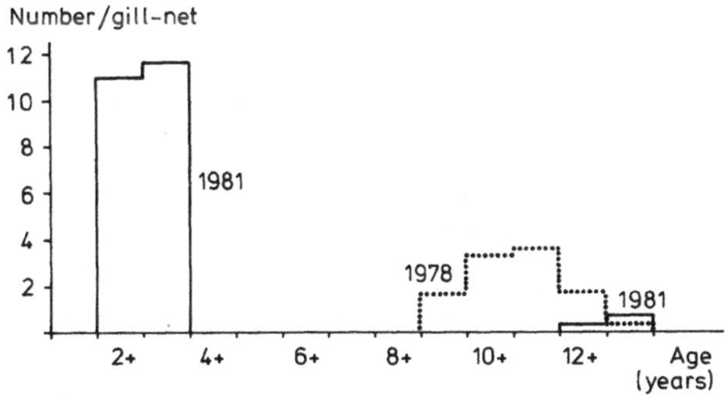

Fig. 14. Age composition of the roach *(Rutilus rutilus)* population in Lake Mörtsjön in 1978 and 1981.

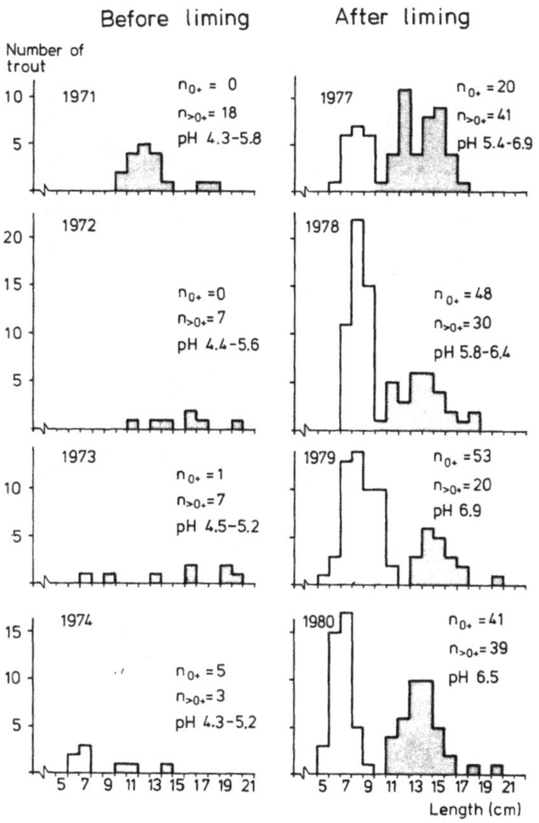

Fig. 15. Catch of young sea trout *(Salmo trutta)* by electro-fishing in River Tjöstelrödsån before and after lime treatment.

Liming before total recruitment failure has in many cases resulted in one or more rich year classes. After liming and fertilizing of Lake St. Rösjön in 1962 a year class of Arctic char *(Salvelinus alpinus)* was produced, which still in 1970 made up 50% of the gill-net catch (Lindström & Andersson 1981).

In a small spawning stream on the Swedish west coast young sea trout *(Salmo trutta)* were almost totally absent during 1971–1974 when pH was in the range of 4.3–5.8. After liming of an upstream lake in 1975 spawning was successful and electrofishing in 1977–1980 resulted in catches of 20–53 one-summer olds and 20–41 older fish (Fig. 15) (Lundh 1981).

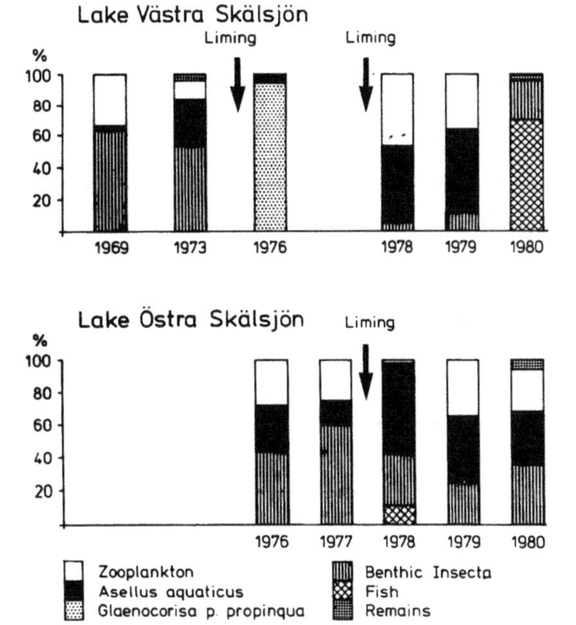

Fig. 16. Diet of older (>15 cm) perch *(Perca fluviatilis)* in lakes V. and Ö. Skälsjön.

Restocking of young Arctic char *(Salvelinus alpinus)* in lakes (Lindström & Andersson 1981; Nyberg, unpubl.) and eggs, yearlings and mature sea-migrating trout *(Salmo trutta)* (Hultberg & Alenäs 1981) in streams after liming has resulted in successful recruitment.

The recovered or improved recruitment after liming may in some waters depend on decreasing hydrogen-ion concentration. In most waters, however, decreasing concentration and toxicity of aluminium (Muniz 1981) seems to be the main reason.

In strongly acidic lakes with weak populations, fish may feed on organisms which do not make up any great part of the diet of a normal population (Mossberg & Nyberg 1976, Almer & Hanson 1980). The perch *(Perca fluviatilis)* population in Lake V. Skälsjön (Fig. 1) was still unaffected by acidification in 1969 and 1973 and zooplankton, *Asellus aquaticus* and benthic insect larvae were the main food items (Fig. 16). In 1976 the population was almost extinct and due to very slight competition for food, perch fed exclusively on *Glaenocorisa p. propinqua*. After liming and successful recruitment, competition for food increased and in 1978 and 1979 the diet of perch was very much the same as before damages caused by acidification. In the downstream lake (Lake Ö. Skälsjön, Fig. 1), liming was performed before any serious effects on the perch population had occurred and the diet was about the same for the entire period (Fig. 16).

The changing composition of the zooplankton and benthic communities and increasing abundances of certain organisms after lime treatment favors fish populations. For example, brown trout *(Salmo trutta)* in Lake Blanksjön preyed upon *Daphnia longispina* and *Cloeon dipterum* and in Lake Bredvatten the introduced *Lymnaea peregra* constituted an important food item for the same species (Hultberg & Andersson 1981). New year classes of perch *(Perca fluviatilis)* and roach *(Rutilus rutilus)* after liming may be heavily preyed upon, as in Lake V. Skälsjön in 1980 (Fig. 16).

In acid lakes with reduced fish populations, individual growth may be good due to a very slight intra-specific competition for food. This was evident in Lake V. Skälsjön, where growth of perch in 1976 was greatly improved compared to seven years earlier (Fig. 17). Liming resulted in a population increase and more intensive competition for food so that in 1980 individual growth had declined and was approaching pre-acid conditions. Lake St. Skarsjön on the Swedish west coast had a declining population of perch and test fishing in 1971 only yielded 3.5 perch/gill-net. After liming in May 1975 some extremely rich year classes resulted in a very dense population with slow individual growth. Test fishing in 1981 yielded 30 perch/gill-net most of them 4–6 years old and with a maximum size of 15 cm. Apparently high survival, and as a consequence a dense population, caused shortage of food and stunted growth (Fig. 18).

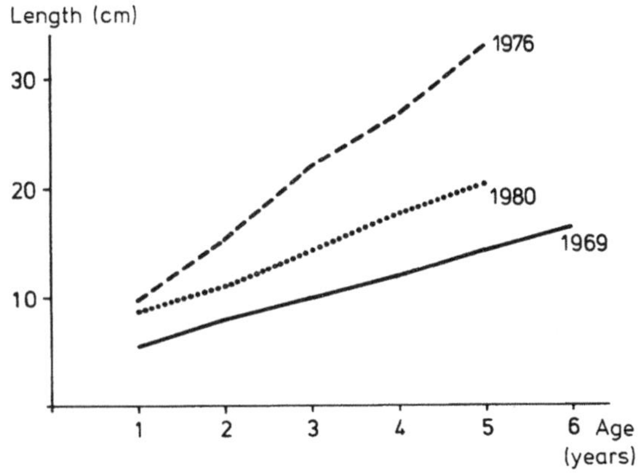

Fig. 17. Growth of perch *(Perca fluviatilis)* in Lake V. Skälsjön in 1969, 1976 and 1980.

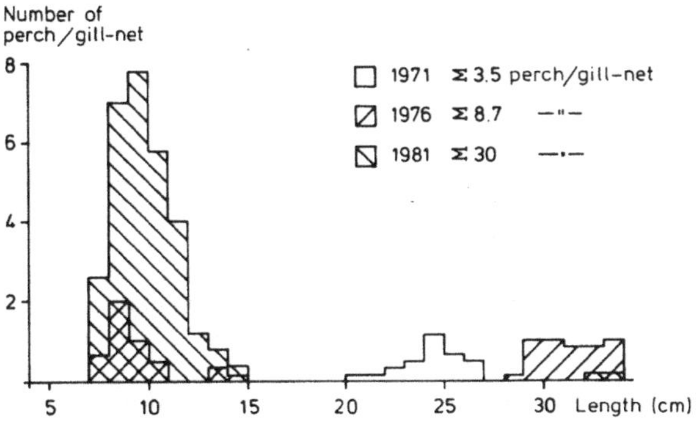

Fig. 18. Size distribution and catch of perch *(Perca fluviatilis)* in Lake St. Skarsjön in 1971, 1976 and 1981.

Growth may be slow even in acidic lakes because of the lack of suitable food organisms, above all fish prey species (Muniz 1981). Growth of perch and pike *(Esox lucius)* seem to be improved after liming resulting in recruitment of roach *(Rutilus rutilus).*

Production of organic material has not been studied at any trophic level. However, the changes noted in the pelagic and benthic communities indicated an increasing production of fish food organisms and should lead to an increased fish production. The yield in numbers of Arctic char *(Salvelinus alpinus)* in Lake Ö. Skälsjön has about doubled after lime treatment and since growth is the same as previously, one might state that production of char has increased. Hultberg & Alenäs (1981) have also found about twice as high densities of young sea trout *(Salmo trutta)* in only slightly acid parts of River Anråseån after liming. The reason for this might be improved survival because of better water quality, but Friberg *et al.* (1980) found more invertebrates in leaf packs in a non-acid than in an acid stream and Andersson (in prep.) found that liming favored detritivorous invertebrates in lakes. It therefore seems reasonable to assume that the increasing production of fish was caused by an increasing production of food organisms.

Conclusions

Many organisms are very sensitive to elevated hydrogen-ion concentrations and certain metal cations. These organisms will be directly favored by an increase in pH and the resultant precipitation of metals, such as aluminium, after liming. Many observed biological effects of acidification are, however, effects of altered biological interactions in the ecosystem. Liming, besides being a possible way to protect waters from acidification, offers great opportunities to check and further study effects of different types of interactions between organisms on all trophic levels in an environment where one does not have to take effects of a toxic water quality into consideration.

The results show, so far, that lime application is a possible method for restoring acidified waters if pH value can be kept at a suitable and stable level. This, however, presents great problems in lakes with brownish water and short turnover times and especially in streams and rivers. In River Högvadsån, for example, a nursery river for atlantic salmon *(Salmo salar)* on the Swedish west coast, low pH and high concentrations of aluminium have caused mortality of young salmon despite considerable expenditure on liming measures. The problems are even more pronounced in many small streams in the northern part of Sweden. The water quality in these is toxic to fish during the spring flood. The flow during this period is, however, so high that it seems almost impossible to improve water quality by liming.

Other alarming effects of acidification which cannot be prevented by liming waters, is the possible reduction of plant nutrients (Hultberg & Andersson 1981) due to decreasing leaching from the drainage area (Broberg & Persson 1981) and the increasing solubility and leaching of aluminium (Muniz 1981) to surface waters.

References

Almer, B., Dickson, W., Ekström, C. & Hörnström, E., 1978. Sulfur pollution and the aquatic ecosystem. In: Nriago, J. O. (ed.). Sulfur in the Environment: II. Ecological Impact, pp. 272–311. John Wiley & Sons, New York.

Almer, B. & Hanson, M., 1980. Effects of acidification in west coast lakes of Sweden. Inf. Inst. freshwat. Res., Drottningholm 5, 44 pp. (in Swedish, with English summary).

Andersson, G., in prep. Decomposition of alder leaves in acid lake waters.

Bergquist, B., 1980. Undersökning av bottenfaunaförekomst, samt kvicksilverhalt i bottenfauna och sediment i sjön Ölen, Örebro län, 1976–1980. Före och efter kalkning. Inst. Limnol. Univ. Upps., 14 pp. (in Swedish).

Brinkhurst, R. O. & Jamieson, B. G. M., 1971. Aquatic oligochaeta of the world. Oliver and Boyd, Edinburgh, 860 pp.

Broberg, O. & Persson, G., 1981. Närsalttillgång i försurade sjöar. Fosfor, kväve och organiskt kol i Gårdsjön. Inst. Limnol. Univ. Upps., 55 pp. (in Swedish).

Brundin, L., 1949. Chironomiden und andere Bodentiere der südschwedischen Urgebirgseen. Rep. Inst. freshwat. Res. Drottningholm 30, 914 pp.

Dickson, W., 1980. Properties of acidified waters. In: Drabløs, D. & Tollan, A. (eds.). Proc. Int. Conf. ecol. Impact Acid Precipitation, pp. 75–83. Sandefjord, Norway.

Eriksson, F., Johansson, J. Å., Mossberg, P., Nyberg, P., Olofsson, H. & Ramberg, L., 1974. The structure of the ecosystem in Lake Vitalampa. Klotenprojektet. Rapp. Scripta Limnol. Ups. 370, 126 pp. (in Swedish, with English summary).

Eriksson, F., Hörnström, E., Mossberg, P. & Nyberg, P., in press. Ecological effects of lime treatment. Inf. Inst. freshwat. Res. Drottningholm 6, 96 pp. (in Swedish, with English summary).

Eriksson, M. O. G., Henriksson, L., Nilsson, B.-I., Nyman, G., Oscarson, H. G., Stenson, A. E. & Larsson, K., 1980. Predatory-prey relations important for the biotic changes in acidified lakes. Ambio 9: 248–249.

Fedorenko, A. Y., 1975. Instar and species-specific diets in two species of Chaoborus. Limnol. Oceanogr. 20: 238–249.

Fiskeristyrelsen and Statens Naturvårdsverk, 1981. Liming of lakes and rivers 1977–1981 in Sweden. Inf. Inst. freshwat. Res. Drottningholm 4, 201 pp. (in Swedish, with English summary).

Friberg, F., Otto, C. & Svensson, B. S., 1980. Effects of acidification on the dynamics of allochthonous leaf material and benthic invertebrate communities in running waters. In: Drabløs, D. & Tollan, A. (eds.). Proc. Int. Conf. ecol. Impact Acid Precipitation, pp. 304–305. Sandefjord, Norway.

Grahn, O., 1977. Macrophyte succession in Swedish lakes caused by deposition of airborne acid substances. Wat. Air Soil Pollut. 7: 295–305.

Grahn, O., Hultberg, H. & Landner, L., 1974. Oligotrofication – a self-accelerating process in lakes subjected to excessive supply of acid substances. Ambio 3: 93–94.

Guiset, A., 1977. Stomach contents of Asplanchna and Ploesoma. Arch. Hydrobiol. Beih. 8: 126–129.

Harvey, H. H., Dillon, P. J., Kramer, J. R., Pierce, R. C. & Whelpdale, D. M., 1981. Acidification in the Canadian aquatic environment. Scientific criteria for an assessment of the effects of acidic deposition on aquatic ecosystems. Natn. Res. Council Can., Ottawa, 369 pp.

Hendrey, G. R. & Wright, R. F., 1976. Acid precipitation in Norway: effects on aquatic fauna. J. Great Lakes Res. 2: 192–207.

Henrikson, L., Nyman, H. G., Oscarson, H. G. & Stenson, J. A. E., 1981. Rotatorieutvecklingen i en försurad sjö efter behandling med kalciumhydroxid, $Ca(OH)_2$. Inst. Zool. Univ. Gothenburg, 5 pp. (in Swedish).

Hinneri, S., 1976. On the ecology and phenotypic plasticity of vascular hydrophytes in a sulphate-rich, acidotrophic freshwater reservoir, SW coast of Finland. Ann. bot. fenn. 13: 97–105.

Hobaek, A. & Raddum, G. G., 1980. Zooplankton communities in acidified lakes in south Norway. SNSF-project IR 75/80. Internal Report IR 75/80. SNSF-Project. Ås, Norway, 132 pp.

Hultberg, H. & Alenäs, I., 1981. Projekt Anråseån. Sammanfattning av uppföljande undersökningar efter utförda kalkningar. Swed. Wat. Air Pollut. Res. Inst. Göteborg, 65 pp. (in Swedish).

Hultberg, H. & Andersson, I., 1981. Liming of acidified lakes and streams – aspects on induced physical-chemical and biological changes. IVL Publ. Göteborg, 43 pp.

Hutchinson, G. E., 1975. A treatise on limnology. John Wiley & Sons, New York, 660 pp.

Hörnström, E., 1981. Trophic characterization of lakes by means of qualitative phytoplankton analysis. Limnologica, Berlin 13: 249–261.

Hörnström, E., Ekström, C. & Holm, L., in prep. Acidification effects on phyto- and zooplankton in lakes on the Swedish west coast.

Iversen, J., 1929. Studien über die pH-Verhältnisse dänischer Gewässer und ihren Einfluss auf die Hydrophyten-Vegetation. Bot. Tidsskr. 40: 277–326.

Jansson, M., 1981. Induction of high phosphatase activity by aluminum in acid lakes. Arch. Hydrobiol. 93: 32–44.

Johansson, K. & Nyberg, P., 1981. Acidification of surface waters in Sweden – effects and extent 1980. Inf. Inst. freshwat. Res. Drottningholm 6, 118 pp. (in Swedish, with English summary).

Kjellberg, G., 1972. Autekologiska studier över Leptophlebia vespertina i en mindre skogstjärn 1966–1968. Ent. Tidskr. 93: 1–29 (in Swedish, with English summary).

Leivestad, H., Hendrey, G., Muniz, I. P. & Snekvik, E., 1976. Effects of acid precipitation on freshwater organisms. In: Braekke, F. (ed.). Impact of Acid Precipitation on Forest and Freshwater Ecosystems in Norway. Research Report FR 6/76, pp. 87–111. SNSF-Project. Ås, Norway.

Lindström, T. & Andersson, G., 1981. Population ecology of salmonid populations on the verge of extinction in acid environments. Rep. Inst. freshwat. Res. Drottningholm 59: 81–96.

Lohammar, G., 1938. Wasserchemie und höhere Vegetation Schwedischer Seen. Symb. Bot. Ups. 3: 252.

Lohammar, G., 1965. The vegetation of Swedish lakes. Acta phytogeogr. suec. 50: 28–48.

Lundh, I., 1981. Effects of liming on the sea trout population in River Tjöstelrödsån. Inf. Inst. freshwat. Res. Drottningholm 7: 24 (in Swedish, with English summary).

Macan, T. T., 1977. Changes in the vegetation of a moorland fishpond in twenty-one years. J. Ecol. 65: 95–106.

Moore, J. W., 1975. The role of algae in the diet of Asellus aquaticus L. and Gammarus pulex L. J. anim. Ecol. 44: 719–730.

Mossberg, P., 1979. Benthos of oligotrophic and acid lakes. Inf. Inst. freshwat. Res. Drottningholm 11: 40 (in Swedish, with English summary).

Mossberg, P. & Nyberg, P., 1976. Effects of acidification on bottom fauna and fish in Lake Västra Skälsjön. Inf. Inst. freshwat. Res. Drottningholm 9: 23 (in Swedish, with English summary).

Mossberg, P. & Nyberg, P., 1979. Bottom fauna of small acid forest lakes. Rep. Inst. freshwat. Res. Drottningholm 58: 77–87.

Muniz, I. P., 1981. Acidification – effects on aquatic organisms. In: Faxxolare, R. A. & Smith, C. B. (eds.). Beyond the

Energy Crisis. Opportunity and Challange, pp. 101–123. Pergamon Press, New York.

Muniz, I. P. & Leivestad, H., 1981. Acidification – effects on freshwater fish. In: Drabløs, D. & Tollan, A. (eds.). Proc. Int. ecol. Impact Acid Precipitation, pp. 84–92. Sandefjord, Norway.

Pejler, B., 1957. Taxonomical and ecological studies on planktic Rotatoria from central Sweden. K. svenska Vetensk-Akad. Handl. 6: 52.

Pourriot, R., 1977. Food and feeding habits of Rotifera. Arch. Hydrobiol. Beih. 8: 243–260.

Raddum, G. G. & Saether, O. A., 1980. Chironomid communities in Norwegian lakes with different degree of acidification. Verh. int. Ver. Limnol. 21: 399–405.

Rosén, G., 1981. 1 000 lakes. The water quality requirements of phytoplankton. Nan. Swed. Envir. Protec. Bd, SNV Rapp., 119 pp. (in Swedish, with English summary).

Saether, O. A., 1979. Chironomid communities as water quality indicators. Holarct. Ecol. 2: 65–74.

Sand-Jensen, K. & Rasmussen, L., 1978. Macrophytes and chemistry of acidic streams from lignite mining areas. Bot. Tidsskr. 72: 105–112.

Scheider, W. & Dillon, P. J., 1976. Neutralization and fertilization of acidified lakes near Sudbury, Ontario. Proc. 11th Can. Symp.: Wat. Pollut. Res. Can., Ontario: 93–100.

Sevalrud, I. H., Muniz, I. P. & Kalvenes, S., 1980. Loss of fish populations in southern Norway. Dynamics and magnitude of the problem. In: Drabløs, D. & Tollan, A. (eds.). Proc. Int. Conf. ecol. Impact Acid Precipitation, pp. 350–351. Sandefjord, Norway.

Statens Naturvårdsverk, 1981. Kodlista SO. Vattenundersökningar. SNV PM. 1492, 277 pp. (in Swedish).

Sörensen, H., 1948. Studies on the ecology of Danish water- and bog-mosses. Dansk bot. Ark. 12: 19–47.

Traaen, T. S., 1980. Effects of acidity on decomposition of organic matter in aquatic environments. In: Drabløs, D. & Tollan, A. (eds.). Proc. Int. Conf. ecol. Impact Acid Precipitation, pp. 340–341. Sandefjord, Norway.

Wiederholm, T. & Eriksson, L., 1977. Benthos of an acid lake. Oikos 29: 261–267.

Okland, J., 1980. Environment and snails (Gastropoda): Studies of 1 000 lakes in Norway. In: Drabløs, D. & Tollan, A. (eds.). Proc. Int. Conf. ecol. Impact Acid Precipitation, pp. 322–323. Sandefjord, Norway.

Okland, K. A., 1980. Mussels and crustaceans: Studies of 1 000 lakes in Norway. In: Drabløs, D. & Tollan, A. (eds.). Proc. Int. Conf. ecol. Impact Acid Precipitation, pp. 324–325. Sandefjord, Norway.

GPSR Compliance

The European Union's (EU) General Product Safety Regulation (GPSR) is a set of rules that requires consumer products to be safe and our obligations to ensure this.

If you have any concerns about our products, you can contact us on ProductSafety@springernature.com

In case Publisher is established outside the EU, the EU authorized representative is:

Springer Nature Customer Service Center GmbH
Europaplatz 3
69115 Heidelberg, Germany

Batch number: 10798364

Printed by Printforce, the Netherlands